4th International Symposium on Functional Materials

(ISFM 2011)

Journal of Physics: Conference Series Volume 339

Sendai, Japan
2-6 August 2011

Editors:

Shu Yin
Tohru Sekino
Shun-ichiro Tanaka
Tsugio Sato
Li Lu
Dongfeng Xue

ISBN: 978-1-62276-685-7
ISSN: 1742-6588

Printed from e-media with permission by:

Curran Associates, Inc.
57 Morehouse Lane
Red Hook, NY 12571

Some format issues inherent in the e-media version may also appear in this print version.

Printed by Curran Associates, Inc. (2013)

For permission requests, please contact the Institute of Physics
at the address below.

Institute of Physics
Dirac House, Temple Back
Bristol BS1 6BE UK

Phone: 44 1 17 929 7481
Fax: 44 1 17 920 0979

techtracking@iop.org

Additional copies of this publication are available from:

Curran Associates, Inc.
57 Morehouse Lane
Red Hook, NY 12571 USA
Phone: 845-758-0400
Fax: 845-758-2634
Email: curran@proceedings.com
Web: www.proceedings.com

TABLE OF CONTENTS

012001 Synthesis of Nano-sized Hydroxyapatite by Ultrasound Irradiation ... 1
Y. Kojima, K. Kitazawa, N. Nishimiya

012002 Synthesis and in Vitro Evaluation of Hydroxyapatite with Controlled Morphology ... 5
M. Kamitakahara, T. Saito, K. Ioku

012003 Preparation and Performance of a Sulfur/Graphene Composite for Rechargeable Lithium-Sulfur Battery ... 9
F. Zhang, Y. Dong, Y. Huang, G. Huang, X. Zhang, L. Wang

012004 Preparation of α-Fe_2O_3 Particles with Controlled Shape and Size Via a Facile Hydrothermal Route ... 14
Q. Dong, S. Yin, C. S. Guo, H. H. Li, N. Kumada, T. Takei, Y. Yonesaki, N. Kinomura, T. Sato

012005 Effects of Sb Addition on ZnO Grain Growth and the Electrical Characteristics of Ba-added-Bi-based ZnO Varistors ... 18
A. Fukumori, A. Kubota, Y. Sato, S. Yoshikado

012006 Sintering and Mechanical Properties of Gadolinium-doped Ceria Ceramics ... 24
K. Yasuda, K. Uemura, T. Shiota

012007 Effects of Sb, Zr, and Y Addition on the Electrical Characteristics of Bi-based ZnO Varistors ... 28
T. Watanabe, Y. Tokoro, Y. Sato, S. Yoshikado

012008 Synthesis and Characterization of Poly(sodium Acrylate)/ Bentonite Superabsorbent Composite ... 33
Y. Xie, Y. Wei, Y. Huang, J. Lin, J. Wu

012009 Effect of pH on Formation of γ-MnO_2 from Zn-bearing $MnSO_4$ Solution ... 37
C. X. Liu, L. Z. Zhang, L. Xiang

012010 Application of Doped Rare-Earth Oxide TiO_2:(Tm^{3+}, Yb^{3+}) in Dye-sensitized Solar Cells ... 43
G. Xie, Y. Wei, L. Fan, J. Wu

012011 Thermal Degradation Kinetics of Poly {N-[(4-bromo- 3,5-difluorine)-phenyl]maleimide-co-styrene} in Nitrogen ... 47
Y. Li, L. Xu, X. Yao, T. Luo, G. Liu

012012 Structural Analysis of TiO_2 and TiO_2-Ag Thin Films and Their Antibacterial Behaviors ... 55
J. H. Hsieh, R. B. Yu, Y. K. Chang, C. Li

012013 Effect of Transition Metal Elements Addition on the Properties of Nitrogen-doped TiO_2 Photocatalysts ... 60
H. Li, P. Zhang, S. Yin, Y. Wang, Q. Dong, C. Guo, T. Sato

Author Index

Preface

The 4th International Symposium on Functional Materials (ISFM2011) was held in Sendai, Japan, on 2–6 August 2011. This Special Issue of *Journal of Physics: Conference Series (JPCS)* consists of partial manuscripts which were presented at ISFM2011.

Advanced materials have experienced a dramatic increase in demand for research, development and applications. The aim of the International Symposium on Functional Materials (ISFM) was to provide an overview of the present status with historical background and to foresee future trends in the field of functional materials. The 4th symposium, ISFM 2011, covered a wide variety of topics within state-of-the-art advanced materials science and technology, and focused especially on four major categories including: Environmental Materials, Electronic Materials, Energy Materials and Biomedical Materials.

As you know, a massive earthquake and the Tsunami that followed occurred near the Tohoku region on 11 March 2011. After the earthquake, although there were many difficulties in continuing to organize the symposium, we received warm encouragement from many researchers and societies, especially from the members of the International Advisory Committee and Organizing Committee, so that ISFM2011 could be held on schedule. We are honored that ISFM2011 was the first formal international academic conference held in the Tohoku area of Japan after the 11 March earthquake.

About 140 participants from 14 countries took part in the ISFM2011 symposium, which included five plenary talks by world-leading scientists, 32 invited talks, and many oral and poster presentations. We are delighted to see that many researchers are interested in the synthesis and the properties as well as the applications of functional materials. Many fruitful and exciting research achievements were presented in the symposium. We believe that this symposium provided a good chance for scientists to communicate and exchange opinions with each other. We would also like to express our sincere appreciation to all the members of the International Advisory Committee, the Organizing Committee, and all the authors and participants. It is expected that the published output of this special issue will be accepted as an original and valuable contribution to the literature in the functional materials field.

Guest Editors
Dr. Shu Yin, Tohoku University, Japan
Dr. Tohru Sekino, Tohoku University, Japan
Prof. Shun-ichiro, Tanaka, IMRAM, Tohoku University, Japan
Prof. Tsugio Sato, IMRAM, Tohoku University, Japan
Prof. Li LU, National University of Singapore, Singapore
Prof. Dongfeng Xue, Changchun Institute of Applied Chemistry, Chinese Academy of Sciences, China

Group photograph of the participants of ISFM2011 held in Sendai, August 4th 2011.

All papers published in this volume of *Journal of Physics: Conference Series* have been peer reviewed through processes administered by the proceedings Editors. Reviews were conducted by expert referees to the professional and scientific standards expected of a proceedings journal published by IOP Publishing.

Synthesis of nano-sized hydroxyapatite by ultrasound irradiation

Y Kojima*, K Kitazawa and N Nishimiya
Department of Materials and Applied Chemistry, Faculty of Science and Engineering, Nihon University, Kanda-Surugadai 1-8, Chiyoda-ku, Tokyo 101-8308, Japan

*E-mail: ykojima@chem.cst.nihon-u.ac.jp

Received 27 July 2011, Revised 20 November 2011

Abstract. Ultrasound irradiation is an effective means of synthesizing fine particles by promoting contact between ions in a solution. We previously reported the synthesis of nano-sized amorphous calcium carbonate particles (specific surface area: 70 m^2/g) using this method. The present study focused on the production of nano-sized hydroxyapatite (HAp) particles with a high specific surface area using ultrasound irradiation. The initial Ca/P atomic ratio was adjusted from 1.00 to 2.00 and the effects on the resulting HAp particles was investigated. HAp particles were found to be easily formed within 5 min of the start of the reaction, and were confirmed to have a single phase using X-ray diffraction analysis. The maximum specific surface area of the particles was approximately 130 m^2/g.

1. Introduction

Ultrasound corresponds to elastic waves with frequencies at and above 20 kHz. When ultrasound irradiation is applied to liquid phase chemical reactions, it has the same effect as high-speed stirring. For example, a frequency of 20 kHz corresponds to 20000 vibrations (oscillations)/s. This results in an increase in the number of contacts between ions in the liquid, which can facilitate the synthesis of extremely fine particles[1-3)]. We previously reported the synthesis of a fine calcium carbonate compound by ultrasound irradiation during an aqueous reaction[4,5)]. Precipitation of calcium carbonate particles was achieved by bubbling CO_2 through a quicklime suspension. The particles had a specific surface area of 100 m^2/g. However, in the case of calcium sulfate dihydrate (gypsum dihydrate), for which the precipitation process had a non-zero induction period, it was found to be extremely difficult to form fine particles in the presence of ultrasound irradiation. However, the technique was successful in producing fine particles of dicalcium phosphate dihydrate ($CaHPO_4 \cdot 2H_2O$; DCPD) from a solution of $Ca(H_2PO_4)_2$, which is the only water-soluble phosphate. Upon continued irradiation of the DCPD suspension, a transformation was found to occur from DCPD to hydroxyapatite (HAp). This implies that ultrasound irradiation is capable of promoting dissolution-precipitation in addition to merely increasing the ion contact frequency. This study investigated the synthesis of hydroxyapatite with high specific surface area by ultrasound irradiation.

2. Experimental

A 0.1 – 0.2 mol/dm^3 $CaCl_2$ solution was first irradiated with ultrasound (Branson, Sonifier 250D (200W) or 2000bdc 40 (400 W)) at a frequency of 20 or 40 kHz, and an amplitude of 123 or 39 μm, respectively. A 0.1 mol/dm^3 Na_3PO_4 solution was rapidly added and ultrasound irradiation was continued. Ca/P atomic ratios of 1.00, 1.50, 1.67 and 2.00 were used and the reaction time was in the range 3 – 20 min. The final sample was obtained by filtrating the suspension and heating at 70 °C.

The samples were characterized using X-ray diffraction (XRD; Rigaku, MultiFlex), infrared (IR) absorption spectroscopy (Shimazu, FTIR-8400s), thermal analysis (TG-DTA; Rigaku, Thermo Plus) , inductively coupled plasma (ICP; SII, SPS1200A) and field emission scanning electron microscopy (FE-SEM; Hitachi, s-4500).

3. Results and discussion

The temperature of the solutions was found to increase as a result of ultrasound irradiation. The 70% amplitude of 20 kHz and 40 kHz is 86 μm and 27 μm, respectively. Figure 1 shows the temperature of a suspension as a function of irradiation time at 20 and 40 kHz. In the case of 20 kHz, it increased to 80 °C during the first 10 min and reached 90 °C after 20 min. Moreover, in the case of 40 kHz, higher temperatures were achieved compared with 20 kHz, because the number of oscillations per second of the ultrasonic horn was larger. HAp is synthesized at temperature more than 80 °C.

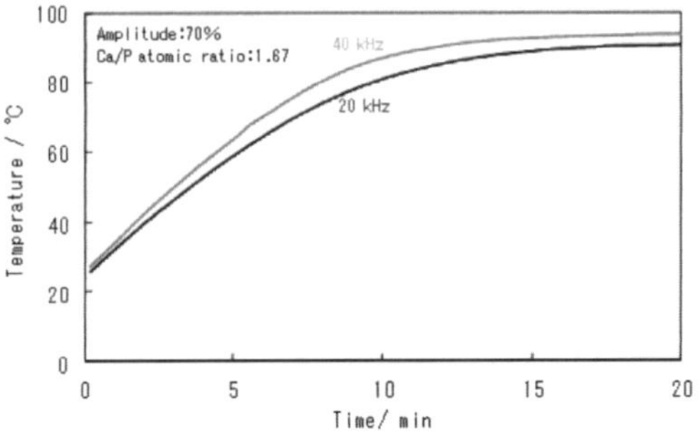

Figure 1 Change in temperature of suspension by ultrasound irradiation.

3.1 Synthesis of HAp by ultrasound irradiation at 20 kHz

Precipitation occurred rapidly when the Na_3PO_4 solution was added to the $CaCl_2$ solution. The formation of precipitates was also confirmed by changes in the electric conductivity of the solution. The pH of the solution was monitored during the reaction. For a Ca/P atomic ratio of 1.00, the pH following the reaction was found to be 12, which was the same as that of the initial Na_3PO_4 solution. For other Ca/P atomic ratios, the pH was observed to decrease immediately after the reaction. Following formation of the deposit, the pH of the suspension declined rapidly and then remained constant in the range 3 - 5.

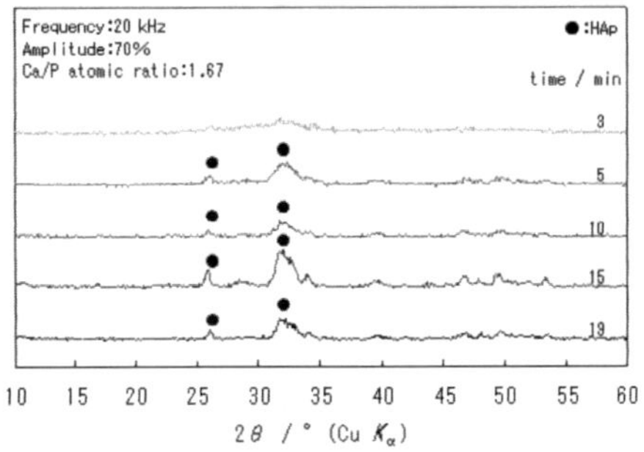

Figure 2 XRD patterns of products synthesized by ultrasound (20 kHz) irradiation.

Figure 2 shows XRD patterns of powder specimens produced at a Ca/P atomic ratio of 1.67 for different reaction times. After 3 min, a halo can be observed at 2θ=30 – 35°, suggesting the formation of amorphous calcium phosphate (ACP) [6),7)]. After 5 min, the characteristic diffraction pattern associated with crystalline HAp was observed. For reaction times above 10 min, the crystallinity of the HAp continued to increase, although little further improvement occurred after 15 min. This sample was subjected to differential thermal analysis, and an exothermic peak due to crystallization of ACP was only weakly observed at about 700 °C.[8)] However, an endothermic peak due to dehydration of ACP was observed at approximately 100 – 200 °C. The above results seem to suggest that ACP is produced in the initial stage of the reaction.

In addition, the HAp was evaluated using IR absorption spectroscopy. In the case of commercial HAp, three absorptions appear at positions close to 600 cm^{-1}. The HAp obtained in the present study for different synthesis times was found to exhibit two additional absorptions at about 1100 cm^{-1}, which are associated with PO_4^{3-}, thus indicating that the HAp is Ca^{2+} deficient.

Figure 3 shows the chemical compositions of the samples products obtained in the present study. Figure 3(a) shows the Ca/P atomic ratio and Fig. 3(b) shows the Na/Ca atomic ratio. These compositions were the average value of the sample which were dissolved with nitric acid, and were measured with ICP. It can be seen that for all initial Ca/P atomic ratios, the Ca/P atomic ratio of the final HAp product was always in the range 1.3 – 1.5, again indicating that it is Ca^{2+} deficient. On the other hand, the Na/Ca atomic ratio for the HAp synthesized with an initial Ca/P atomic ratio of 1.00 decreased with increasing reaction time from 0.4 to 0.2. This value is very high and suggests the formation of a $NaCaPO_4$ precursor.

Figure 4 shows the effect of the initial Ca/P atomic ratio and the reaction time on the specific surface area of the HAp product. The specific surface area was highest for initial Ca/P atomic ratio of 1.00. However, since it is believed that this HAp was mixed together with the $NaCaPO_4$ precursor, this result was excluded from the analysis. For initial Ca/P atomic ratio of 2.00, the specific surface area of the HAp increased during the first 10 min, reaching a value of approximately 130 m^2/g, because fine HAp particles were being formed by crystallization of ACP. It then decreased as a result of HAp crystal growth by a thermally driven dissolution-precipitation mechanism (see Fig. 1).

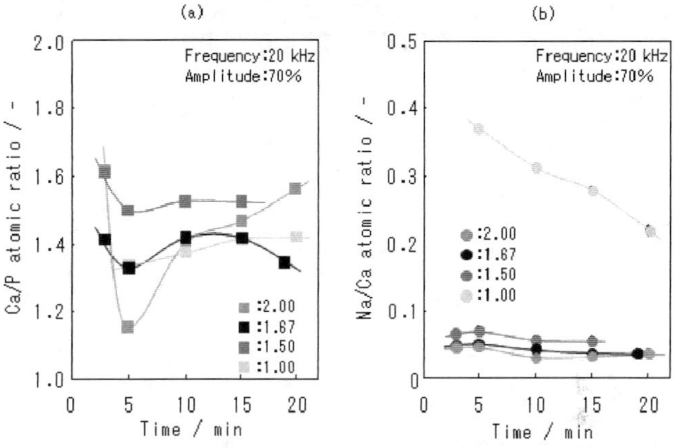

Figure 3 Chemical compositions of HAp products obtained by ultrasound irradiation.

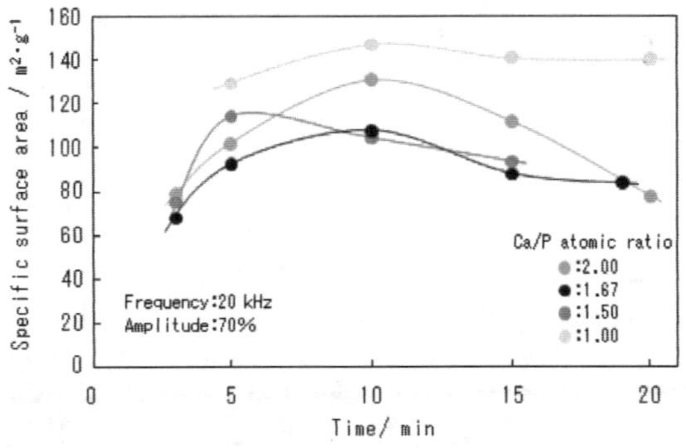

Figure 4 Effect of initial Ca/P atomic ratio on specific surface area of HAp obtained by ultrasound irradiation.

Figure 5 FE-SEM micrographs of HAp products obtained by ultrasound irradiation.

Figure 5 shows FE-SEM micrographs of HAp particles produced with different Ca/P atomic ratios, following a reaction time of 10 min. The HAp appears as spindle-like crystals with sizes of 30 – 50 nm.

3.2 Synthesis of HAp by ultrasound irradiation at 40 kHz

Next, the ultrasound frequency was changed to 40 kHz. At this frequency, the horn oscillates more rapidly but the oscillation amplitude decreases.

Figure 6 shows XRD patterns of products synthesize

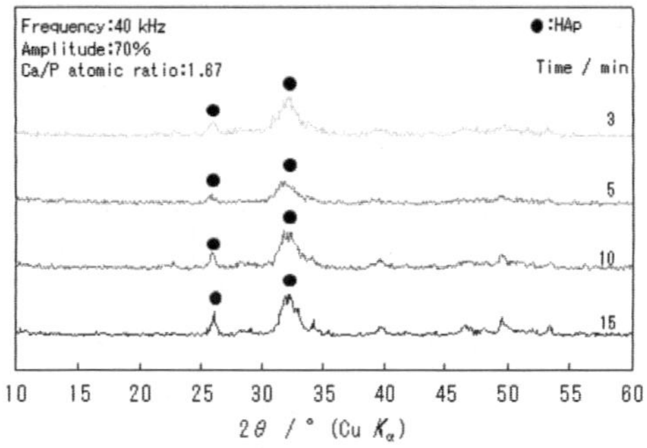

Figure 6 XRD patterns of HAp synthesized by ultrasound (40 kHz) irradiation.

ratio of 1.67. As discussed in Section 3.1, for a frequency of 20 kHz, a halo-like pattern was observed after 3 min (Fig. 2). However, at a frequency of 40 kHz, the characteristic pattern of HAp was already observed for a reaction time of 3 min, and the crystallinity increased with increasing reaction time.

Figure 7 shows the dependence of the specific surface area of the products on the ultrasound frequency and reaction time. As a reference, the specific surface area of HAp which was synthesized by stirring at 80 °C was approximately 100 m^2/g. It can be seen that higher specific surface areas were obtained at 40 kHz than at 20 kHz. This is due to the increased oscillation frequency of the horn, which accelerates the transformation from ACP to HAp. The maximum specific surface area for HAp particles synthesized at 40 kHz was 115 m^2/g for a reaction time of 5 min.

Figure 8 shows FE-SEM micrographs of HAp particles synthesized at 40 kHz. The particles have sizes of about 30 nm and a spindle shape. In the case of the reference HAp specimen synthesized by stirring, the particles were also spindle shaped but their average size was about 50 nm.

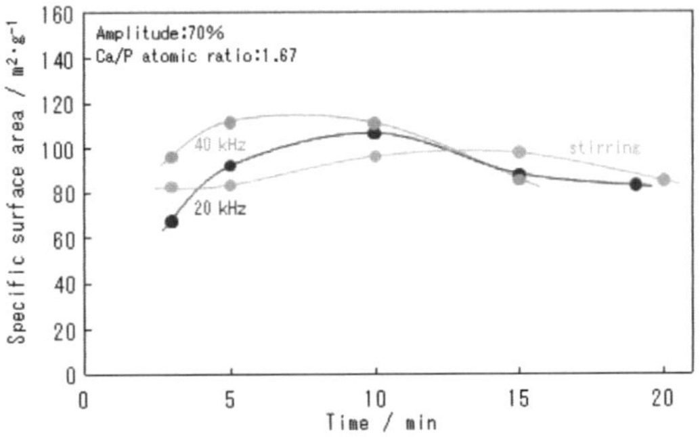

Figure 7 Effect of initial Ca/P atomic ratio on specific surface area of HAp obtained by ultrasound irradiation.

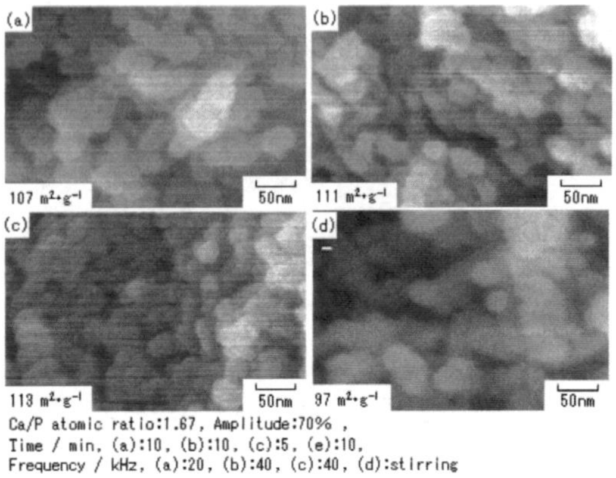

Figure 8 FE-SEM micrographs of HAp products obtained by ultrasound irradiation.

4. Summary

This study investigated the synthesis of fine HAp particles in a liquid-phase reaction under ultrasound irradiation. At an ultrasound frequency of 20 kHz, ACP was formed in the initial stage of the reaction, and Ca^{2+}-deficent HAp was then formed by crystallization of ACP. The specific surface area of HAp synthesized at a Ca/P atomic ratio of 2.00 for 10 minutes was 130 m^2/g and the average particle size was determined to be 30 – 50 nm. At 40 kHz, HAp particles formed more rapidly than at 20 kHz, and their sizes were smaller. Thus, ultrasound irradiation is capable of producing fine HAp particles with specific surface areas of over 100 m^2/g and is therefore a very promising method for biomaterials and ion-exchange material

Reference

1) M.A.Alavi,A.Morsali,Ultrasonics Sonochem.,15,833-838(2008).
2) W.Jiang,X.Gong,Z.Chen,Y.Hu,X.Zhang,X.Gong,Ultrasonic Sonochem., 14,208-212 (2007).
3) F.Dang,K.Kato,H.Imai,S.Wada,H.Haneda,M.Kuwabara,Ultrasonic Sonochem.,17,310-314(2010).
4) Y.Kojima,K.Yamaguchi, N.Nishimiya,Ultrasonic Sonochem.,17,617-620(2010).
5) Y.Kojima, K.Yamaguchi, N.Nishimiya,J.Soc.Inogr.Mater.Japan,16,171-175(2009).
6) T.Yasue, K.Masuda, Y.Kojima,J.Ceram.Soc.Japan,102,1122-1127(1994).
7) Y.Kojima,K.Sakama,T.Yasue,Y.Arai,Phos.Res.Bull.,4,47-52(1994).

Synthesis and *in vitro* evaluation of hydroxyapatite with controlled morphology

M Kamitakahara*, T Saito and K Ioku
Graduate School of Environmental Studies, Tohoku University
6-6-20 Aoba, Aramaki, Aoba-ku, Sendai, 980-8579, Japan

*E-mail: kamitaka@mail.kankyo.tohoku.ac.jp

Received 11 August 2011, Revised 16 November 2011

Abstract. Hydroxyapatite (HA) ceramics have been used as bone-repairing materials. Particles in HA ceramics show different properties depending on their compositions and morphologies. It is expected that the control of the composition and morphology contributes to improvement of biological properties of HA ceramics. In order to reveal the effect of the composition and morphology, the HA ceramics composed of rod-shaped or plate-shaped particles with calcium-deficient composition were prepared by the hydrothermal process. The *in vitro* evaluation, simulated body fluid (SBF) soaking test, was then conducted for these samples. The calcium-deficient HA ceramics composed of rod-shaped or plate-shaped particles were reacted with SBF and incorporated calcium and phosphate ions from SBF. This implies that these HA ceramics show high affinity to bone in the body.

1. Introduction

The technology which repairs the damaged bone is important because the damage in bones decreases the quality of life of the patient. Some ceramics are used for repairing the damaged bone. Hydroxyapatite (HA; $Ca_{10}(PO_4)_6(OH)_2$) is a main inorganic component of bone, and HA ceramics are focused as bone-repairing materials. HA ceramics can bond to bone directly and sintered HA ceramics have been clinically used as bone-repairing materials [1]. However, the little resorption [2] and low ability to stimulate bone formation of the sintered HA ceramics are problems. As the properties of particles in HA ceramics depend on their composition and morphology, it is expected that controlling the morphology and composition of HA particles would contribute to an improvement in the biological properties of HA ceramics. We previously prepared calcium-deficient HA ceramics composed of rod-shaped particles, which exposed the *a*-face preferentially, through a hydrothermal process. These ceramics stimulated bone regeneration and were resorbed completely in rabbit femora [3,4]. This regeneration is likely due to the calcium-deficiency of the composition and the exposure of a specific face of the HA crystal. Bioresorbable calcium-deficient HA could be expected to contribute to the development of new biodegradable bone substitutes.

However, the knowledge about the effect of the composition and morphology of the HA has been insufficient and it is needed to reveal why calcium-deficient HA ceramics composed of rod-shaped particles show superior properties. In the present study, calcium-deficient HA ceramics composed of rod-shaped or plate-shaped particles were prepared by the hydrothermal process, and *in vitro* evaluation, simulated body fluid (SBF) soaking test [5], was conducted.

2. Materials and methods

2-1 Preparation of samples

The commercial available alpha-tricalcium phosphate (α-TCP, $Ca_3(PO_4)_2$, Taihei Chemical Industrial Co., Ltd. Japan) powder 0.10 g was pressed into a pellet 12 mm in diameter by a uniaxial pressing. Three pellets were immersed in 30 cm^3 of an ammonia solution, whose pH was controlled at 11, in a Teflon®-lined sealed vessel. The vessel was kept at 160 °C for 24 h for the hydrothermal treatment. The sample was named Rod-HA.

In order to obtain the pellets composed of plate-shaped HA particles, octacalcium phosphate (OCP, $Ca_8H_2(PO_4)_6{\cdot}5H_2O$) was used as a starting material. The OCP particles were obtained by the reaction of phosphoric acid and calcium carbonate in distilled water at 60 °C according to the method previously reported [6]. The obtained OCP powder 0.10 g was pressed into a pellet 12 mm in diameter by a uniaxial pressing. Three pellets were immersed in 30 cm^3 of a nitric acid solution, whose pH was controlled at 4, in a Teflon®-lined sealed vessel. The vessel was kept at 160 °C for 24 h for the hydrothermal treatment. The sample was named Plate-HA.

The crystal phase and surface morphology of the obtained samples were examined by X-ray diffraction (XRD, RINT-2200VL, Rigaku Co. Japan) and scanning electron microscopy (SEM, S-4100, Hitachi Ltd. Japan), respectively. The Ca/P molar ratio of the products was determined based on a previously developed method [7,8]. The resultant HA particles were heated at 900 °C for 3 h to decompose the samples into a mixture of stoichiometric HA and β-TCP. We calculated the Ca/P molar ratio by determining the ratio of HA and β-TCP formed in the decomposition by XRD and then applying the calibration curve of HA and β-TCP [7].

2-2 In vitro test

To estimate the chemical reaction of the obtained HA samples in the body, the obtained pellets were soaked in a simulated body fluid (SBF) which has ion concentrations nearly equal to those of human blood plasma [5]. The ion concentrations are shown in **Table 1**. This solution does not contain proteins and cells, and is widely used to reveal the chemical reaction between the materials and the body fluid. The pellets were soaked in SBF with pH 7.40 for 3 d, and then removed from the SBF. The changes in the pellet surfaces were examined by SEM. The changes in the element concentrations of SBF were examined by inductively coupled plasma atomic emission spectroscopy (ICP-AES, iCAP 6200, Thermo Scientific Inc, USA).

Table 1. Ion concentrations of human blood plasma and simulated body fluid (SBF)

	Concentration / $mol{\cdot}m^{-3}$							
	Na^+	K^+	Mg^{2+}	Ca^{2+}	Cl^-	HCO_3^-	HPO_4^{2-}	SO_4^{2-}
Blood plasma	142.0	5.0	1.5	2.5	103.0	27.0	1.0	0.5
SBF	142.0	5.0	1.5	2.5	147.8	4.2	1.0	0.5

3. Results and Discussion

Figures 1 and 2 show the XRD patterns and SEM photographs of the obtained pellets. The XRD patterns indicated that the crystal phase of both the pellets was HA. However, the morphologies of the particles which construct the pellets were different. For the sample Rod-HA which was prepared from α-TCP, rod-shaped HA particles were observed. It has been reported that the hydrothermal treatment of α-TCP provides the rod-shaped HA particles [9]. On the other hand, for the sample Plate-HA which was prepared from OCP, plate-shaped HA particles were observed. The morphology of the starting material, OCP particle, was plate-shaped and this plate-shaped morphology was maintained even after the hydrothermal treatment. This result is consistent with the previous study [10], in which plate-shaped HA was obtained by the hydrothermal treatment of OCP particles. The Ca/P molar ratios of Plate-HA and Rod-HA were 1.61 and 1.62, respectively. This indicates that the obtained HA samples have calcium-deficient compositions.

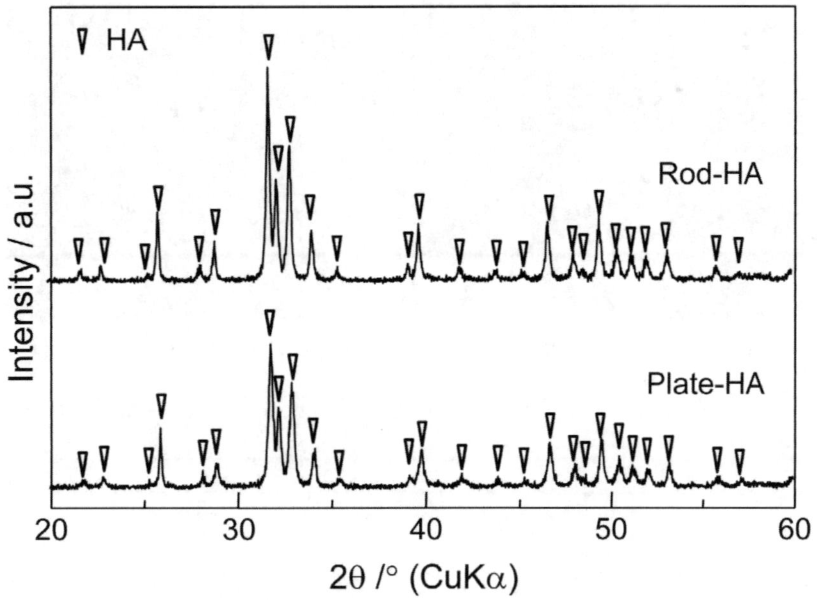

Figure 1. XRD patterns of the pellets.

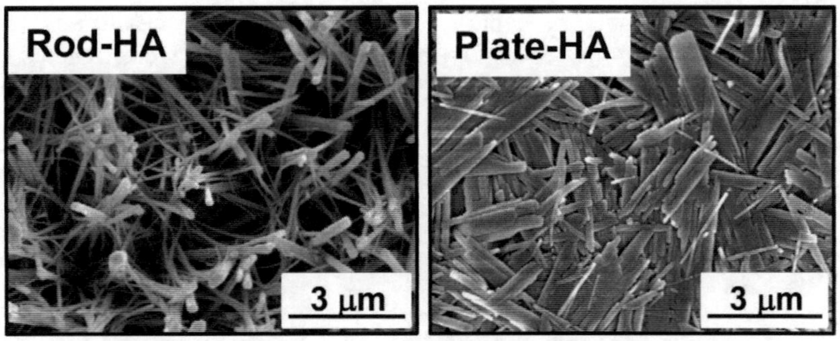

Figure 2. SEM photographs of the surfaces of the pellets.

Figure 3 shows the SEM photographs of the pellets after soaking in SBF for 3 d. The changes by soaking in SBF were not observed from SEM observation under this magnification. However, the calcium (Ca) and phosphorus (P) concentrations of SBF were significantly decreased after soaking the Rod-HA and Plate-HA pellets. The Ca and P concentrations of SBF after soaking the pellets for 3 d are shown in **Table 2**. These imply that Ca-deficient HA ceramics composed of rod-shaped or plate-shaped particles were reacted with SBF and incorporated calcium and phosphate ions from SBF. It is speculated that the HA particles prepared by the hydrothermal treatment were highly reactive with SBF and small precipitates or crystal growth of the original HA particles occurred. It has been reported that the materials which bond to bone can form the HA layer on their surfaces in the body and bond to bone through this HA layer. This HA formation on the materials can be reproduced in SBF [5]. Therefore, the high reactivity of calcium-deficient HA ceramics composed of rod-shaped or plate-shaped particles implies that these ceramics show high affinity to bone in the body. The high reactivity of the calcium-deficient HA ceramics composed of rod-shaped particles in the body might be one of the reasons why the ceramics show superior properties in body.

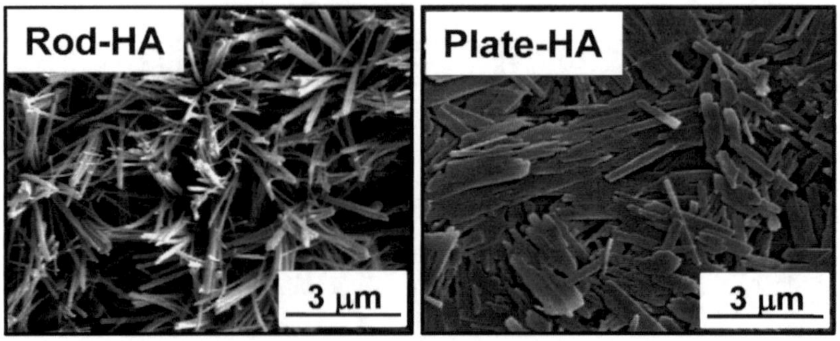

Figure 3. SEM photographs of the surfaces of the pellets after soaking in SBF for 3 d.

Table 2. Ca and P concentrations of SBF after soaking the pellets for 3 d

	Concentration / mol·m^{-3}	
	Ca	P
Original SBF	2.50	1.00
Rod-HA	1.33	0.23
Plate-HA	1.01	0.20

4. Summary

Pellets composed of HA particles with controlled morphology and composition were prepared, and they were soaked in SBF to evaluate the behaviors under physiological conditions. Calcium-deficient HA ceramics composed of rod-shaped or plate-shaped particles were reacted with SBF and incorporated calcium and phosphate ions from SBF. This implies that these ceramics show high affinity to bone in the body.

5. Acknowledgement

This research was partially supported by JSPS KAKENHI (21300175). We are grateful for the experimental help of Profs. K. Tohji and N. Tsuchiya of Tohoku University.

6. References

[1] Hench L L 1991 *J. Am. Ceram. Soc.* **74** 1487
[2] Hoogendoorn H A, Renooij W, Akkermans L M, Visser W and Wittebol P 1984 *Clin. Orthop. Relat. Res.* **187** 281
[3] Okuda T, Ioku K, Yonezawa I, Minagi H, Gonda Y, Kawachi G, Kamitakahara M, Shibata Y, Murayama H, Kurosawa H and Ikeda T 2008 *Biomaterials* **29** 2719
[4] Gonda Y, Ioku K, Shibata Y, Okuda T, Kawachi G, Kamitakahara M, Murayama H, Hideshima K, Kamihira S, Yonezawa I, Kurosawa H and Ikeda T 2009 *Biomaterials* **30** 4390
[5] Kokubo T and Takadama H 2006 *Biomaterials* **27** 2907
[6] Kamitakahara M, Okano H, Tanihara M and Ohtsuki C 2008 *J. Ceram. Soc. Japan* **116** 481
[7] Ioku K, Murakami T, Ikuma Y and Yoshimura M 1992 *J. Ceram. Soc. Japan* **100** 1015 [in Japanese].
[8] Ishikawa K, Duchene P and Radin S 1993 *J. Mater. Sci.: Mater. Med.* **4** 165
[9] Ioku K, Kawachi G, Sasaki S, Fujimori H and Goto S 2006 *J. Mater. Sci.* **41** 1341
[10] Kamitakahara M, Ito N, Murakami S, Watanabe N and Ioku K 2009 *J. Ceram. Soc. Japan* **117** 385

Preparation and performance of a sulfur/graphene composite for rechargeable lithium-sulfur battery

Feifei Zhang[1,2], Yunhui Dong[1], Yun Huang[2], Gang Huang[2], Xinbo Zhang[2] and Limin Wang[2,*]

[1] School of Chemical Engineering, Shandong University of Technology, Zibo 255091, China
[2] State Key Laboratory of Rare Earth Resource Utilization, Changchun Institute of Applied Chemistry, Chinese Academy of Sciences, Changchun 130022, China

*E-mail: lmwang@ciac.jl.cn

Received 29 July 2011, Revised 18 November 2011

Abstract. The lithium-sulfur (Li-S) battery is a promising electrochemical system that has high theoretical capacity. The sulfur/graphene nanosheets (S/GNS) composite is prepared through thermal reduction between the sulfur (S) and graphene oxide (GO). The morphology and composition of the composite are analyzed by means of x-ray powder diffraction (XRD), scanning electron microscopy (SEM) and energy dispersive spectroscopic (EDS) characterization. It is found that the element S distributed uniformly between the layers of GNS. Graphene with a two-dimensional structure of carbon atoms is employed as a conductive and absorbing agent for the S cathode materials of rechargeable Li-S battery. The S/GNS composite cathode shows a specific electrochemical capacity, which is about 1598 mAh g^{-1} S at the initial discharge and remains above 670 mAh g^{-1} after 80 cycles.

1. Introduction

Facing the rapid growing energy shortage and global environment problems, the development of Li-ion batteries with high energy-density is in high demand for extending their applications to storing electricity from renewable sources [1]. One of the most promising candidates for storage devices is the Li-S cell. The theoretical specific capacity of S is 1672 mAh g^{-1}, corresponding to a theoretical specific energy of 2600 Wh kg^{-1} [2]. Furthermore, it possesses obvious advantages of abundant resources, low cost and environmental friendliness.

However, the low electrical conductivity of S and S-containing organic compounds limit active material utilization due to poor electrochemical contacts within the material. Moreover, the dissolution of electrochemical reaction products of polysulfides (Li_2S_x, where $1 \leq x \leq 8$) and the redox shuttle mechanism make capacity faded quickly [3]. To enable operating the Li-S battery, the S cathode material must be well combined with a high conductive and a strong adsorbing agent. Various carbon materials, such as active carbon [4], mesporous carbon [5], and carbon nanotubes [6], have been intensively investigated as additives for Li-S batteries.

Recently, GNS, a two-dimensional structure of carbon atoms with high surface area, superior electrical conductivity, chemical stability, and mechanical properties, has attracted high attention for use of energy storage applications [7]. In a recent report by Wang et al., S was melted and coated on the GNS, which shows an improved capacity [8]. Furthermore, in another report, researchers synthesized a graphene-sulfur composite material by synthesizing submicrometer sulfur particles coated with PEG containing surfactants and GNS. The composite showed high specific capacity with relatively good cycling stability as the cathode for Li-S batteries [9].

Herein, we used a facile and low-cost methodology to the large-scale synthesis of S/GNS composite, wherein the GO was used instead of GNS which is easy to be prepared, and S as a reducing agent will be reduce GO to GNS through thermal process. Meanwhile, melted S can be homogeneously distributed between the layers of GNS to take advantage of the fact that S interacts strongly with carbon. The lyophilized method can also keep a high surface area of the composite.

2. Experimental

2.1. Materials Synthesis and characterization

The GO was made by a modified Hummers method [10], and then mixed with elemental S (Aladdin Reagent, AR) in the weight ratio of 1:5, with the mixture designated S/GO. The S/GO was ultrasonicated for 30 min to get a uniform suspension and then lyophilized it. The mixture was held at 423 K for 4 h under argon gas protection to allow the melted elemental S to infiltrate into the layers of GNS. Then, the temperature was increased to 573 K and held for 2 h, reduced GO to GNS.

The microstructure analyses were characterized using XRD measurements being performed on a Rigaku-Dmax 2500 diffractometer with Cu Kα radiation. SEM was performed on a field emission Hitachi S-4800 instrument.

2.2. Electrochemical measurements

The working cathode slurry was prepared by mixing 80 wt.% composite with 10 wt.% acetylene black (Hong-xin Chemical Works) and 10 wt.% polyvinylidene fluoride (PVDF, DuPont Company, 99.9%) binder in N-methyl-2- pyrrolidinone (NMP, Aladdin Reagent, AR) solvent. A sulfur cathode slurry containing 50 wt. % S, 40 wt.% carbon black, and 10 wt.% PVDF binder was also prepared in the same way to compare with the working cathode. The slurries were coated onto Al foil current collector using doctor's blade. It was dried in a vacuum oven at 333 K for 24 h and then pressed. Subsequently, the electrode was cut into disks with a diameter of 12 mm. Coin-type cells of 2025 were assembled in an Ar-filled glove box with H_2O and O_2 content below 1 ppm, with Li foil as anode, 1 M $LiPF_6$ EC/DMC (v/v, 1:1) as electrolyte. The charge/discharge performances were tested between 1.2 V and 3 V, using LAND CT2001A multi-channel battery testing system at room temperature. The electrochemical impedance response (EIS) measurements were carried out with a BioLogic VMP3 station.

3. Results and discussion

3.1. The structure characterization

Figure 1. SEM images of (a) elemental S powder, (b) GNS and (c) S/GNS composite.

Figure 1 shows the SEM images of element S, GNS and their composites prepared through reduction. The bare S power consists of microsized particles (Figure 1a). Figure 1b shows the GNS reduced by decomposition at 573 K for 4 h under argon gas protection. After heat treatment of S/GO, the microsized S particles melted and coated uniformly on the GNS sheets (Figure 1c).

To verify the composition of our S/GNS composite, we carry out EDS mapping/imaging of the material. Chemical mapping confirms that the bright particles in the SEM image (Figure 2a) are element C (Figure 2c, carbon mapping), with element S being coated on the GNS sheets uniformly (Figure 2d, S mapping). EDS spectrum captured for the region shown in Figure 2b also shows the existence of S.

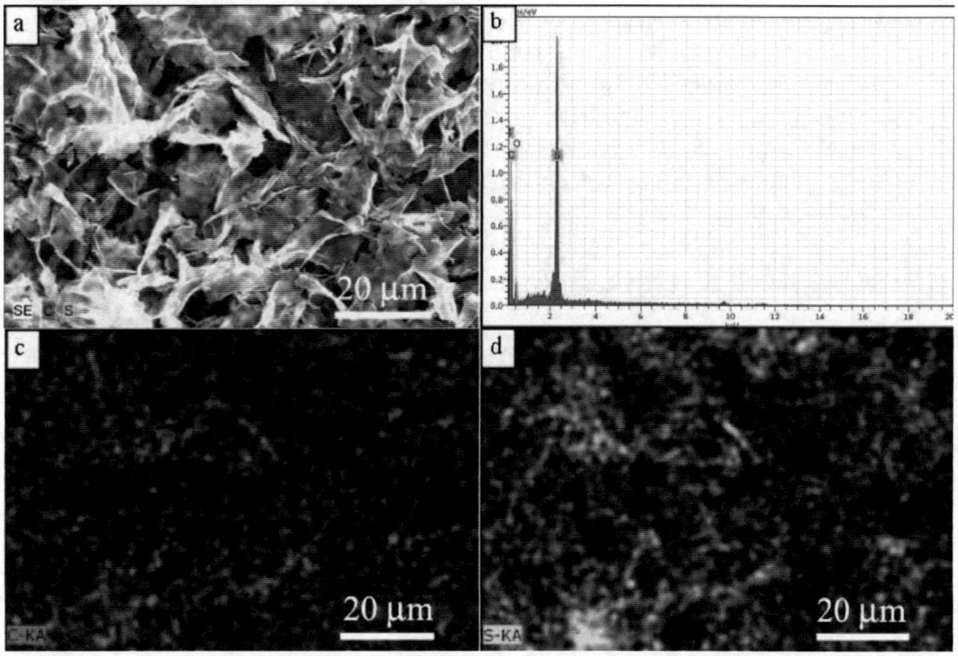

Figure 2. EDS characterization of S/GNS composite. (a) SEM image of GNS coated S particles. (b) EDS spectrum captured for the region shown in (a). (c) EDS mapping of the carbon region shown in (a). (d) EDS mapping of the S region shown in (a).

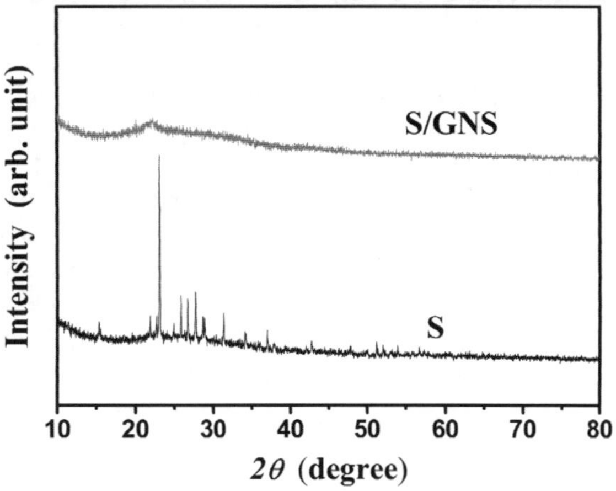

Figure 3. XRD patterns of elemental S and S/GNS composite.

Figure 3 shows the XRD patterns for the S powder and S/GNS composite. Elemental S exhibits several sharp peaks from 10 to 60 degree, indicating its good crystal state. For the S/GNS composite with a low S content of 22 wt.%, the characteristic peaks of crystal S are not detectable, which suggests that S becomes amorphous and homogeneously distributed in the composite.

3.2. Electrochemical performance of the S/GNS

Figure 4 shows the discharge curves of the S and S/GNS composite vs. Li/Li^+. The discharge curves represent a typical characterization of Li-S batteries. The discharge curves show two plateaus based on the voltage profile. The detailed mechanisms for oxidation and reduction of S, polysulfides and Li sulfide during discharge-charge were already reported. The upper plateau is well known as the change from elemental S to the higher order Li polysulfides (Li_2S_n, $8>n>2$). The lower plateau is caused by the reduction of the higher order Li polysulfides to lower order Li sulfides (Li_2S_2, Li_2S). Calculated from the S content (22 wt.%), the initial discharge specific capacity of the working cathode is about 1598 mAh g^{-1}, the utilization of electrochemically active S is 95.6% assuming a complete reaction to the product of Li_2S. In contrast, the initial discharge specific capacity of the S cathode is about 1094 mAh g^{-1}.

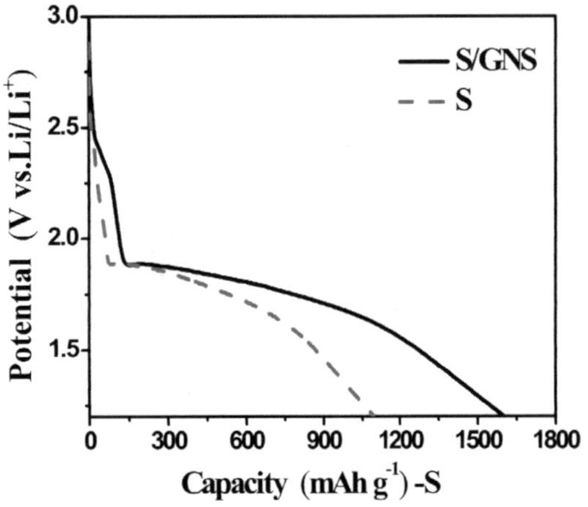

Figure 4. First discharge curves of the S and S/GNS composite electrodes

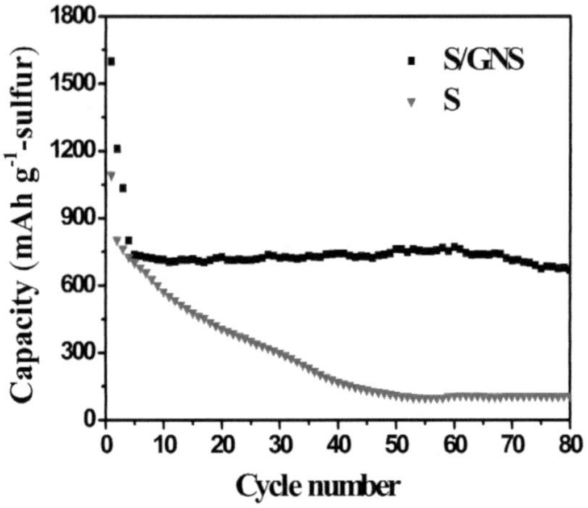

Figure 5. Discharge capacities vs cycle number for S and S/GNS composite electrodes

The cycling performance of the S/GNS composite electrode is compared to that of the pure S, as shown in Figure 5, being cycled at a constant current density of 160 mAh g^{-1} S. The discharge specific capacity of the composite electrode drops from initial 1598 mAh g^{-1} to 670 mAh g^{-1} after 80 cycles, which shows a good cyclic stability. The initial capacity of the S/GNS electrode is higher than that of the pure S electrode and the cycling stability of the S/GNS composite is also improved.

EIS is also invited to investigate the differences of electrochemistry of the two electrodes. Nyquist profiles are illustrated in Figure 6. The simplified equivalent circuit on the upper-right is used to interpret the measured data. It can be see from the figure that all the Nyquist plots of cathodes are composed by a semicircle at high frequencies relating to the contact resistance and charge transfer resistance, and a short inclined line in low frequency regions due to the ion diffusion within the cathode. The semicircle in S/GNS electrode is much smaller than the S electrode, because GNS can provide a better electron and ionic conductivity network.

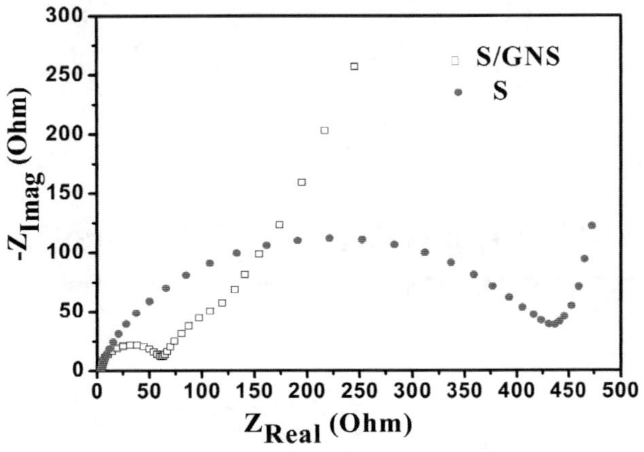

Figure 6. Impedance plots for electrodes of S and S-GNS composite.

4. Conclusions

A S/GNS composite with layered structure is prepared by thermal treatment with a mixture of elemental S and GO, the GO can be well reduced to GNS and elemental S is coated uniformly onto the GNS through heat treatment. The S/GNS composite shows a specific capacity, which is about 1598 mAh g^{-1} S at the initial discharge and remains above 670 mAh g^{-1} even after 80 cycles. GNS can provide a better electron and ionic conductivity network in the composite and can also contain the polysulfides effectively. The S/GNS composite is a promising cathode material for rechargeable Li-S battery.

Acknowledgements

This work is supported by the following funds: (1) Shandong Provincial Natural Science Foundation, China (No. ZR2009BM045); (2) National Nature Science Foundations of China (No. 21073179); and (3) the Foundation for Innovative Research Groups of the National Natural Science Foundation of China (No. 20921002).

References

[1] Ji X L, Lee K T and Nazar L F 2009 *Nat. Mater.* **8** 500–6A
[2] Marmorstein D, Yu T H, Striebel K A, McLarnon F R, Hou J and Cairns EJ 2000 *J. Power Sources.* **89** 219-26
[3] Jayaprakash N, Shen J, Moganty S S, Corona A and Archer L A 2011 *Angew. Chem. Int. Ed.* **50** 1-6
[4] Wang J L, Yang J, Xie J Y, Xu N X and Li Y 2002 *Electrochem. Commun.* **4** 499-502
[5] Liang X, Wen Z, Liu Y, Zhang H, Huang L and Jin J 2011 *J. Power Sources.* **196** 3655-8
[6] Yuan L, Yuan H, Qiu X, Chen L and Zhu W 2009 *J. Power Sources.* **189** 1141-6
[7] Worsley M A, Pauzauskie P J, Olson T Y, Biener J, Satcher J H, Jr and Baumann T F 2010 *J. AM. CHEM. SOC.* **132** 14067-9
[8] Wang J Z, Lu L, Choucair M, Stride J A, Xu X and Liu H K 2011 *J. Power Sources.* **196** 7030-4
[9] Wang H, Yang Y, Liang Y, Robinson J T, Li Y, Jackson A, Cui Y and Dai H 2011 *Nano Lett.* **11**, 2644–7
[10] Marcano D C, Kosynkin D V, Berlin J M, Sinitskii A, Sun Z, Slesarev A, Alemany L B, Lu W and Tour J M 2010 *ACSNANO.* **4** 4806-14

Preparation of α-Fe_2O_3 particles with controlled shape and size via a facile hydrothermal route

Q Dong[1,2*], S Yin[1], C S Guo[1], H H Li[1], N Kumada[2], T Takei[2], Y Yonesaki[2], N Kinomura[2] and T Sato[1]

[1]Institute of Multidisciplinary Research for Advanced Materials, Tohoku University, 2-1-1, Katahira, Aoba-ku, Sendai 980-8577, Japan

[2]Department of Research Interdisciplinary Graduate School of Medicine and Engineering, University of Yamanashi, Miyamae cho-7, Kofu 400-8511, Japan

*E-mail: dong@mail.tagen.tohoku.ac.jp

Received 22 October 2011, Revised 29 November 2011

Abstract. α-Fe_2O_3 particles with controlled shape and size have been prepared by a facile hydrothermal reaction between ferric chloride hexahydrate ($FeCl_3{\cdot}6H_2O$) and sodium hydroxide (NaOH) at 160°C. X-ray powder diffraction pattern indicates that the product is single-phase of α-Fe_2O_3.The morphology of α-Fe_2O_3 particles changed as homogenous sphere, top-shape and cube, depending on solution pH. Meanwhile, monodispersed cube particles of α-Fe_2O_3 are also prepared under the same conditions using surfactant of hexadecyl trimethyl ammonium bromide (CTAB). The experimental results suggest that particle size can be controlled by adjusting the solution pH and amount of CTAB.

1. Introduction

The synthesis of α-Fe_2O_3 particles with different shapes and sizes were intensively investigated because of their novel physical and chemical properties compared to the bulk materials, and their potential applications in catalysts, magnetic recording media, pigments, anti corrosive agents and gas sensors, etc. [1-5]. Various techniques have been developed to prepare α-Fe_2O_3 particles, such as sol-gel process, microemulsion technique, forced hydrolysis method, chemical oxidation in polymer or mineral matrixes, electrochemical synthesis and chemical precipitation [6-20]. In these synthesis methods, shape and size of the products change depending on the synthesis parameters, such as concentration of the reactants, temperature, medium, etc. Moreover, multiple reaction steps are required to obtain the final products. By far, hydrothermal method has been widely adopted due to effective control of the size and shape of the nanoparticles, high homogeneity, well-crystallized product and relatively low reaction temperature. Until now, many researchers have prepared α-Fe_2O_3 nanoparticles by hydrothermal approaches [21-23]. To our knowledge, many studies have been done on the preparation of α-Fe_2O_3, which belongs to hexagonal crystal system, by the hydrothermal reaction without additives or with organic additives, such as surfactants and citrate [24-27]. For instance, Kandori et al. have successfully prepared cube particles with uniform micropores of 0.8 nm in diameter by using $FeCl_3$ aqueous solution [24]. Chen et al. have prepared hollow-structured α-Fe_2O_3 particles from a $FeCl_3{\cdot}6H_2O$ and NaOH solution in the presence of DBS (sodium dodecyl benzenesulfonate) [25]. Using $(NH_4)_3Fe(C_2O_4)_3$ and NaOH as starting materials, Jing et al. have synthesized monodispersed ellipsoidal and spherical shapes α-Fe_2O_3 nanoparticle modified by sodium dodecylsulfonate (SDS) and hexadecylpyridinium chloride (HPC), respectively [26]. Diamandescu et al. have synthesized acicular polycrystalline α-Fe_2O_3 particles by using $FeCl_3 \cdot 6H_2O$ and citrate [27]. However, there are few reports concerning the synthesis of α-Fe_2O_3 particles using CTAB (hexadecyl trimethyl ammonium bromide). On the basis of our previous studies [28-31], in this paper, we tried to utilize a facile hydrothermal route to synthesize α-Fe_2O_3 particles with controlled shape and size from a solution of $FeCl_3 \cdot 6H_2O$ and NaOH by controlling the starting pH value of the solution and the amount of CTAB.

2. Experimental

2.1 Synthesis All reagents were of analytical grade and used without further purification. 8 ml $FeCl_3 \cdot 6H_2O$ solution (1.00 mol dm^{-3}) was added to 8 ml NaOH aqueous solution with different concentrations at room temperature under vigorous stirring for 15 min, and the pH values of slurry solutions were fixed to 3, 7, and 11, respectively. Then, the slurry solutions were poured into Teflon-lined autoclaves and hydrothermally heated at 160°C for 12 h. To investigate how the CTAB influences the final products shape and size, different amounts of CTAB were added to the slurry solution with the pH value of 11. After that, the autoclaves were cooled to room temperature, and the products were collected by centrifugation, washed with deionized distilled water and ethanol. Finally, the samples were dried in air at 60°C for 20 h before characterization.

2.2 Characterization X-ray powder diffraction (XRD) was performed on a Rigaku X-ray diffractometer (RINT2000V Japan) with graphite-monochromatized CuKα radiation (λ=1.54056 Å) using 30 mA and 40 kV. The diffraction pattern over the range of 20-60 in 2θ was recorded with a scanning speed of 3°/min. Scanning electron microscopy (SEM) images were taken with JEOL JEM 6500F field emission scanning electron microscope.

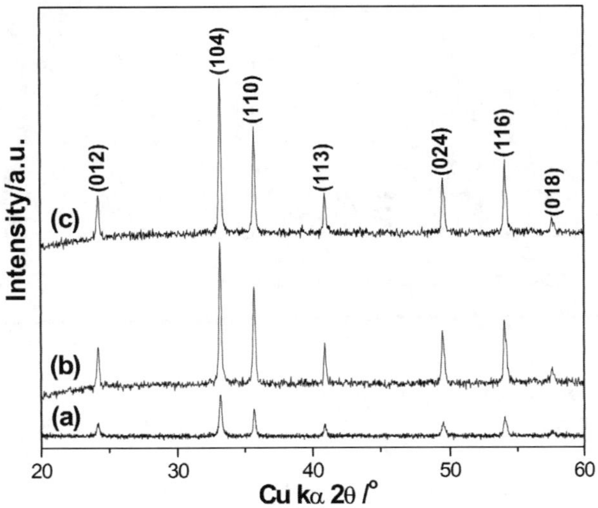

Figure 1. XRD patterns of α-Fe_2O_3 particles prepared at 160°C for 12 h with different pH values without CTAB (a) pH 3, (b) pH 7, (c) pH 11.

3. Results and Discussion

Figure 1 shows the X-ray patterns of the samples prepared at 160°C for 12 h at different pH values, such as 3, 7 and 11. All the patterns exhibit the characteristic XRD pattern of hematite (JCPDS 33-664). The intensity of the diffraction peaks increased with pH value. It is also apparent that the width at half-maximum of the diffraction peaks decreased slightly with increasing pH value. From these results, the crystallinity and particle size of the products increased with pH value. In the other words, higher pH value enhances the growth of the α-Fe_2O_3 [32]. Figure 2 shows the SEM images of the samples. Spherical, top-shaped, cube α-Fe_2O_3 particles can be obtained as pH values are 3, 7 and 11, respectively. The corresponding particle sizes are about 1, 1.5 and 2 μm, respectively. This trend is consistent with the result from XRD. The morphology of α-Fe_2O_3 particles changes together with the particle size; i.e., small particle has spherical form, medium particle is top-shaped, and large particle is cube. The results suggest that the particle shape can be controlled by adjusting pH value in the starting solution.

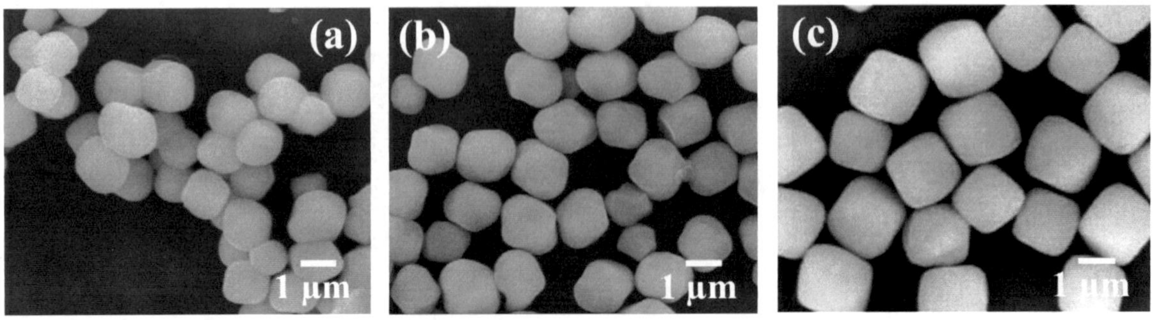

Figure 2. SEM images of α-Fe_2O_3 particles prepared at 160°C for 12 h with different pH values without CTAB additive at (a) pH 3 (b) pH 7 (c) pH 11.

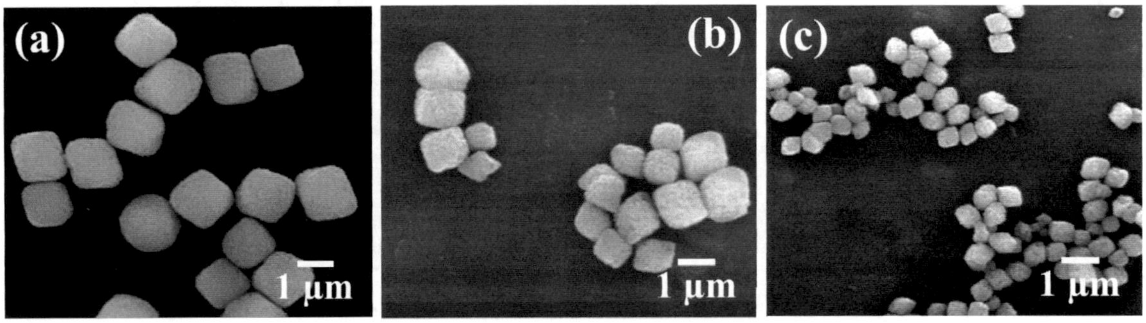

Figure 3. SEM images of α-Fe_2O_3 particles prepared at 160°C for 12 h with (a) 0.25g, (b) 0.5g, (c) 1g CTAB additive at pH 11.

CTAB is a cationic surfactant which ionizes completely in water. Recently, CTAB assisted hydrothermal technique has emerged as an attractive technique to investigate the controlled synthesis of various materials [33-35]. As shown in Fig. 3, almost all particles have cubed shape. Adding 0.25g CTAB (Fig. 3a), cube α-Fe_2O_3 particles with the size about 1 μm was obtained. When the amount of CTAB increased to 0.5g, the size of cube α-Fe_2O_3 particles decreased to about 0.8 μm (Fig. 3b). Cube α-Fe_2O_3 particles have the average size about 0.5 μm (Fig. 3c) as the amount of CTAB increased to 1g. Compared with the particles obtained without adding CTAB (Fig. 2c), the size of the cube α-Fe_2O_3 particles decreases with the amount of CTAB, and the surfaces of the particles become rough (Fig. 3). At the higher amount of CTAB would lead to increase the number of nuclei, therefore, the particle size would decrease considerably [35].

4. Conclusions

α-Fe_2O_3 particles with controlled shape and size have been successfully prepared via a facile hydrothermal method. It is clarified that the starting pH value of the solution plays an important role in the shape of α-Fe_2O_3 particles, and CTAB is responsible to control the size of α-Fe_2O_3 particles. This hydrothermal route can be successfully used for the synthesis of α-Fe_2O_3 particles with controlled shape and size, which are expected to be applied in the related fields, taking the advantage of being an environmentally friendly and less energy consuming procedure.

References

[1] Kodama R H, Makholoufand S A and Berkovitz A E 1997 *Phys. Rev. Lett.* **1393** 79
[2] Jiang J Z, Lin K and Lin W 1997 *J. Phys. D: Appl. Phys.* **30** 1459
[3] Sun H T, Cantalini C and Faccio M 1996 *J. Am. Ceram. Soc.* **79** 927
[4] Ozaki M, Kratchvil S and Matijevic B 1984 *J. Colloid Interface Sci.* **102** 146
[5] Dimitrov D V, Hadjipanayis G C, Papaefthymiou V and Simopoulous A J 1998 *J. Magn. Magn. Mater.* **8** 188
[6] Chunga Y, Lima S K, Kima C K, Kima Y H and Yoona C S 2004 *J. Magn. Magn. Mater.* **272-276** 1167
[7] Pascal C, Pascal J L and Favier F 1999 *Chem. Mater.* **11** 141
[8] León C P, Kador L and Zhang M 2004 *J. Raman Spectrosc.* **35** 165
[9] Itoh H and Sugimoto T 2003 *J.Colloid Interface Sci.* **265** 283.
[10] Liu X Q, Tao S W and Shen Y S 1997 *Sens. Actuators, B, Chem.* **40** 161
[11] Hashimoto T, Yoko T and Saka S 1993 *J. Ceram. Soc. Jpn.* **101** 64
[12] Dong W T and Zhu C S 2002 *J. Mater. Chem.* **12** 1676.
[13] Dong W T, Wu S X and Chen D P 2000 *Chem. Lett.* **5** 496
[14] Ayyub P, Multani M and Barma M 1988 *J. Phys. C. Solid State Phys.* **21** 2229
[15] Hamada S and Matijevic E 1982 *J. Chem. Soc., Faraday Trans.* **178** 2147
[16] Hamada S, Niizrki S and Kudo Y 1986 *Bull. Chem. Soc. Jpn.* **59** 3443
[17] Nakatari Y and Matsuoka M 1982 *Jpn. J. Appl. Phys.* **21** L758
[18] Glisenti A 1998 *J. Chem. Soc., Faraday Trans.* **94** 3671
[19] Vasquez-Mansilla M, Zysler R D and 1999 *J. Magn. Magn. Mater.* **204** 29
[20] Vasquez-Mansilla M, Zysler R D and Arciprete C C 2001 *J. Magn. Magn. Mater.* **226-230** 1907
[21] Li G S, Smith Jr R L, Inomata H and Arai K 2002 *Mater. Res. Bull.* **37** 949
[22] Sahu K K, Rath C, Mishre N C, Anand S and Das R P 1997 *J. Colloid Interface Sci.* **185** 402
[23] Rath C, Sahu K K and Kulkarni S D 1999 *Appl. Phys. Lett.* **75** 4171
[24] Kandori K and Ishikawa T 2004 *J.Colloid Interface Sci.* **272** 246
[25] Chen D H, Jiao X and Zhao Y 2003 *J. Mater. Chem.* **13** 2266
[26] Jing Z and Wu S 2004 *Mater Res Bull.* **39** 2057
[27] Diamandescu L and Tarabasanu D M 1999 *Ceram Inter.* **25** 689
[28] Dong Q, Wang D, Yao J X, Kumada N, Kinomura N, Takei T, Yonesaki Y and Cai Q 2009 *J. Ceram. Soc. Jpn.* **117** 245
[29] Dong Q, Kumada N, Yonesaki Y, Takei T and Kinomura N 2009 *J. Ceram. Soc. Jpn.* **117** 881
[30] Dong Q, Kumada N, Yonesaki Y, Takei T and Kinomura N 2010 *J. Ceram. Soc. Jpn.* **118** 272
[31] Dong Q, Kumada N, Yonesaki Y, Takei T, Kinomura N and Wang D 2010 *J. Mater. Sci.* **45** 5685
[32] Zhang J H, Kong Q H, Dua J, Ma D K, Xi G C and Qian Y T 2007 *J. Crystal. Growth.* **308** 159
[33] Chen L, Zhang D J, Chen J M, Zhou H D and Wan H Q 2006 *Mater. Sci. Eng., A* **415**156
[34] Wang Y X, Sun J, Fan X Y and Yu X 2011 *Ceram Inter.* **37** 3431
[35] He Q J, Cui X Z, Cui F M, Guo L M and Shi J L 2009 *Microporous Mesoporous Mater.* **117** 609

Journal of Physics: Conference Series **339** (2012) 012005 doi:10.1088/1742-6596/339/1/012005

Effects of Sb addition on ZnO grain growth and the electrical characteristics of Ba-added-Bi-based ZnO varistors

A Fukumori, A Kubota, Y Sato and S Yoshikado*
Department of Electronics, Doshisha University, Kyoto 610-0321, Japan

*E-mail:syoshika@mail.doshisha.ac.jp

Received 8 August 2011, Revised 25 November 2011

Abstract. The varistor voltage increases as the number of ZnO grain boundaries between electrodes increases. Therefore, in order to fabricate varistors with low varistor voltages, it is necessary to reduce the number of ZnO grain boundaries between electrodes. The present study uses this method for increasing the ZnO grain size. With the goal of fabricating varistors with low varistor voltages, the effect on the ZnO grain size of adding Sb to Bi- Ba-Co-Mn-added ZnO varistors was investigated. Ba was added in order to increase the ZnO grain size, and Sb was added in order to achieve a uniform ZnO grain size without reducing the grain size. The addition of a small amount of Sb reduced the formation of ZnO grains having a smaller grain size, and the addition of 0.02 mol% Sb to the 0.5-mol%-Bi- and 0.5 mol% Ba added ZnO varistor exhibited a low varistor voltage of approximately 31 V/mm and the highest resistance to electrical degradation, because compounds containing both Ba and Mn do not form at grain boundaries between ZnO grains.

1. Introduction

ZnO varistors with low breakdown voltages (varistor voltages) are used to protect electronic devices with low operating voltages (e.g., control circuits and cellular phones) from external surges and excessive voltages caused by static electricity discharges [1]. ZnO varistors must have highly nonlinear voltage–current (*V–I*) characteristics and a high resistance to electrical degradation. The varistor voltage increases as the number of ZnO grain boundaries between electrodes increases. Therefore, in order to fabricate varistors with low breakdown voltages, it is necessary to reduce the number of ZnO grain boundaries between electrodes. Two methods have been proposed to achieve this goal, namely, reducing the electrode separation and increasing the ZnO grain size. Laminated ZnO varistors used in practical applications are realized by reducing the electrode separation. However, in order to realize still lower varistor voltage using this technique, it is necessary to generate an even thinner layer. However, the resulting decreased yield is problematic.

Adding only Ba or Ti to Bi-based ZnO varistors promotes grain growth, which enables large ZnO grains to be obtained [2]. The present study uses this method to increase the ZnO grain size. Adding only Ba to Bi-Co-Mn-added ZnO varistors enhances ZnO grain growth and the electrical characteristics of ZnO varistors [3]. The ZnO grain size increases upon the addition of Ba and Bi, and optimizing the added amounts allows a minimum varistor voltage of approximately 12 V/mm to be obtained. However, the resistance to electrical degradation was poor due to current concentration in large ZnO grains because the ZnO grain size was not uniform and the *V–I* characteristics were determined by the current flowing in the large grains. Adding only Sb to Bi-based ZnO varistors makes the ZnO grain size more uniform, inhibits ZnO grain growth, and improves the resistance to electrical degradation [4]. In the present study, in order to obtain a uniform ZnO grain size without reducing the grain size, a small amount of Sb was added as an additive. The effects of Sb addition on ZnO grain growth, varistor voltage, electrical characteristics, and resistance to electrical degradation were investigated changing the amount of addition of Bi [3,5]. Co and Mn were also added because these are standard additives for commercial Bi-based ZnO varistors. This addition demonstrates that the ZnO orientation is not the primary factor in improved resistance to electrical degradation [5].

2. Experiments

As additives, MnO_2 (0.5 mol%), Co_3O_4 (0.2 mol%), Bi_2O_3 (0 to 1.0 mol%), $BaCO_3$ (0.5 mol%), and Sb_2O_3 (0 to 0.2 mol%) have been added to ZnO powder (purity: 99.999%) [3], revealing that the growth of ZnO grains of 0.5-mol%-Mn- and 0.6-mol%-Co-added samples can be promoted considerably by adding more than 0.35 mol% Ba [3]. For Ba addition of 0.75 mol% or more, the sample was very fragile. The ZnO grain size could be increased significantly by adding approximately 0.5 mol% Ba and approximately 0.5 mol% Bi, and optimizing the amounts added (approximately 0.5 mol% Ba and approximately 0.4 mol% Bi) allows a minimum varistor voltage [3]. Therefore, the amount of Ba is determined to be 0.5 mol%. These starting materials were wet mixed with ethanol. After drying, the powder mixture was calcined in air at 600°C for 5 h, and after being pulverized the mixture was sintered in air at 1,150°C for 10 h. A sintering time of 10 h was chosen because the growth of large ZnO grains was observed for sintering over 10 h. The sintered disk was polished to a thickness of 0.5 mm, and aluminum electrodes of 2.38 mm in diameter were then attached to the end surfaces. In order to observe the structure of ZnO grain boundaries, the sintered disk surface was etched using a dilute NaOH aqueous solution. The average size of 200 ZnO grains was estimated from optical microscopy images by taking the grain size to be equal to the side length of a square with the same area. The value of the nonlinearity index α for the V–I characteristics was estimated using the equation $V = V_N(I/I_N)^{1/\alpha}$. The varistor voltage is defined as V_N per 1 mm when $I_N = 1\times10^{-3}$ A/cm^2. The V–I characteristics for current densities in the range of approximately 3×10^{-7} to 3×10^{-3} A/cm^2 were obtained. Electrical degradation testing was performed by applying a dc current of 100 mA/cm^2 in air for 1 min. The elementary compositions of the samples were analyzed by scanning electron microscopy (SEM) and energy-dispersive X-ray spectroscopy (EDS) and using a back-scattering electron (BSE) detector. The crystal structures of the samples were analyzed by X-ray diffraction (XRD; Cu K_α).

3. Results and discussion

Figure 1 shows the distribution ρ of ZnO grain sizes estimated from the optical microscopy image of the electrode surface for samples (a), (c), and (d), to which 0.5 mol% Bi and 0, 0.02, and 0.1 mol% Sb were added, and sample (b), to which 0.63 mol% Bi and 0.01 mol% Sb were added. The distribution ρ for the cross sectional surface perpendicular to the electrode surface of the sample was approximately the same as that of the electrode surface. This distribution reveals that the grain size of ZnO for the 0.02-mol%-Sb-added sample becomes broad. Figure 2 shows the relationship between average grain sizes of ZnO for the 0.5-mol%-Bi-added samples and added Sb. The average grain size and the lower limit of ZnO became maximal for an addition of approximately 0.02 mol% Sb. Figure 3 shows the relationship between the average grain sizes of ZnO for samples to which 0, 0.01, and 0.02 mol% Sb were added and the amounts of Bi. The average grain size of ZnO became maximal at approximately 0.5 mol% Bi for the sample to which Sb was not added and at the amount of approximately 0.63 and 0.5 mol% for samples added Sb of 0.01 and 0.02 mol%, respectively. The maximum average grain size of ZnO for 0.01-mol%-Sb-added samples was approximately the same as that for 0.02-mol%-Sb-added samples, and its value was approximately 58 μm. This value was approximately 1.3 times larger that that for samples to which Sb was not added. On the other hand, for samples to which 0.03 mol% or more Sb was added, the maximum value became smaller than that for samples to which 0.02

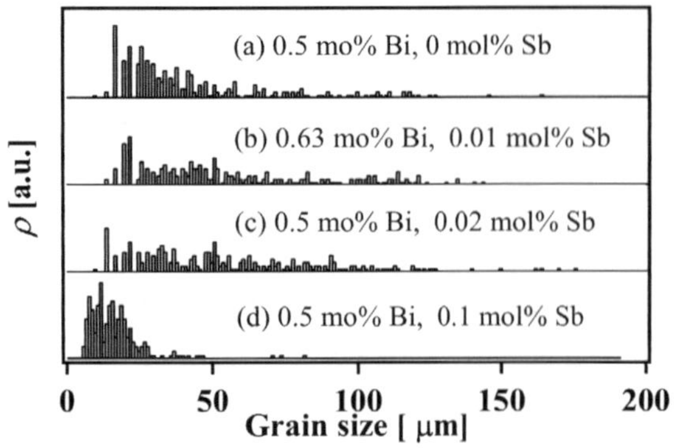

Figure 1. Probability distribution of ZnO grain sizes for samples to which various amounts of Bi and Sb have been added.

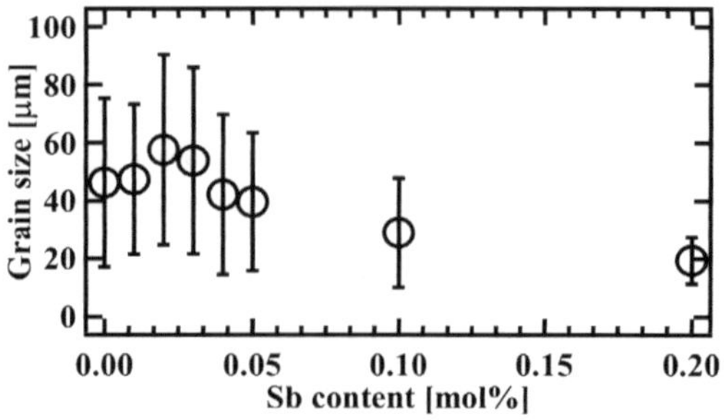

Figure 2. Average grain sizes of ZnO for the samples to which 0.5 mol% Bi and various amounts of Sb have been added.

mol% Sb was added. The grain size of ZnO for both the 0.01-mol%-Sb- and 0.02-mol%-Sb-added samples was approximately the same for the addition of more than 0.63 mol% Bi. The range of the amount of Bi that yields the same average grain size of ZnO for 0.02-mol%-Sb addition became large compared with that for 0.01-mol%-Sb addition.

Figure 4 shows the XRD patterns of various samples. A prominent diffraction peak of ZnO was observed at approximately $2\theta = 31.8°$ for all samples. Diffraction peaks of $Ba_2Mn_3O_8$ and $BaBiO_3$ were observed at approximately $2\theta = 27.2°$ and 31° and at approximately $2\theta = 29.2°$, respectively, for sample (a), to which 0.5 mol% Bi and 0 mol% Sb had been added. Although a prominent diffraction peak of $BaBiO_3$, which promotes the grain growth of ZnO, was not observed for samples (b) and (c), to which Bi was not added, a new unknown peak was observed at approximately $2\theta = 29°$, 31.4°, and 32.2° for sample (c), to which 0.02 mol% Sb was added [5]. A prominent diffraction peak of a new compound at approximately $2\theta = 31.7°$ for samples (d) and (e), to which 0.63 and 0.5 mol% Bi was added, respectively, was observed only when the ZnO grain size was maximal, as shown in Figure 3. Assignment of the compound has never been performed. In contrast, the diffraction peaks for $Ba_3Mn_2O_8$ disappeared. Therefore, the existence of $BaBiO_3$ and the generation of a new compound are

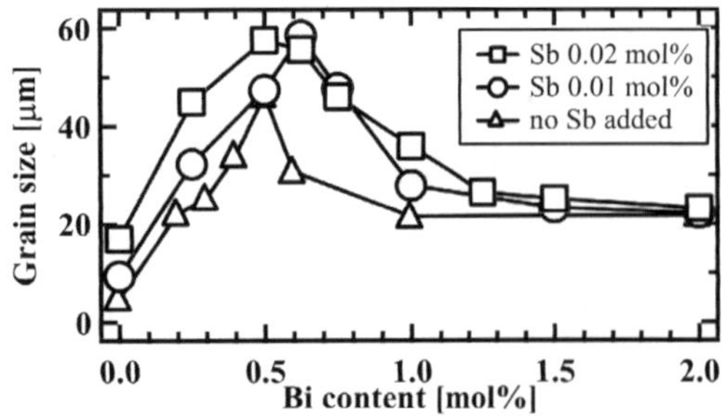

Figure 3. Average grain sizes of ZnO for the samples to which various amounts of Bi and Sb have been added.

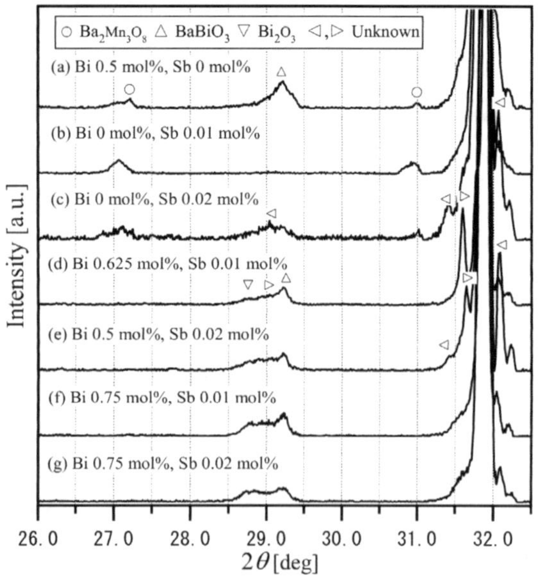

Figure 4. XRD patterns of various samples.

essentially related to the promotion of ZnO grain growth.

Figures 5(a), 5(b), and 5(c) show EDS elementary mappings of Bi, Ba, and Mn, respectively, for the 0.5-mol%-Ba-, 0.5-mol%-Bi-, and 0-mol%-Sb-added samples. Figures 5(d) and 5(e) are BSE images for 0.01-mol%-Sb- and 0.02-mol%-Sb-added samples, respectively. Figures 5(f), 5(g), and 5(h) are elementary mappings of Bi, Ba, and Mn for the 0.01-mol%-Sb-added sample. Figures 5(h), 5(i), and 5(j) are elementary mappings of Bi, Ba, and Mn for 0.02-mol%-Sb-added sample. Figures 5(l), 5(m), and 5(n) show elementary mappings of Ba and Mn for the 0.4-mol%-Sb-added sample. The gray-scale brightness in both the BSE and EDS images increases as the amount and atomic weight of an element increase. For a sample to which Sb was not added, comparing Figure 5(a) with 5(b) also reveals that Ba is segregated at grain boundaries. However, Ba is uniformly distributed at grain boundaries, despite the non-uniform distribution of Bi at grain boundaries. Comparison of Figures 5(b) and 5(c) reveals that compounds or mixtures containing Ba and Mn are segregated at grain boundaries and that Bi is distributed independently of these two elements. The compound in the grain boundaries is identified as $Ba_2Mn_3O_8$ by comparing with the XRD results shown in Figure 4. For 0.01-mol%-Sb- and 0.02-mol%-Sb-added samples, the compound in the grain boundaries is identified as $BaBiO_3$ by comparing the BSE images shown in Figures 5(d) and 5(e) with Figures 5(f), 5(g), 5(i), and 5(j) and the XRD results shown in Figure 4. Due to the formation of $Ba_3Mn_2O_8$, ZnO

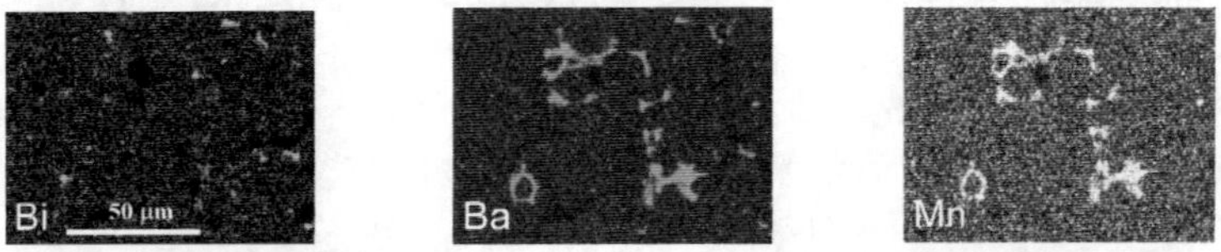

(a) Bi (no Sb added) (b) Ba (no Sb added) (c) Mn (no Sb added)

(d) BSE image (0.01 mol% Sb) (e) BSE image (0.02 mol% Sb)

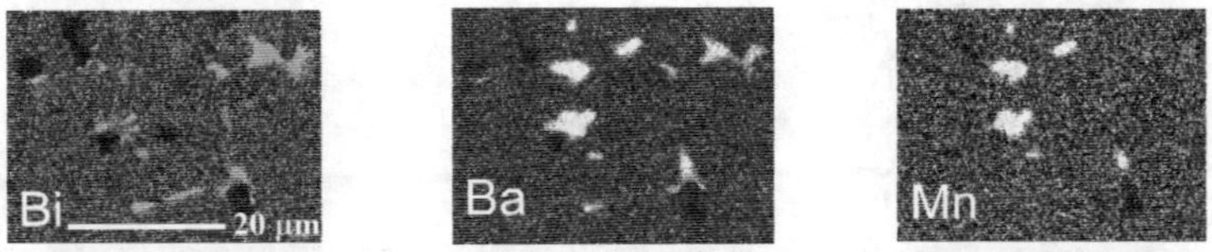

(f) Bi (0.01 mol% Sb) (g) Ba (0.01 mol% Sb) (h) Mn (0.01 mol% Sb)

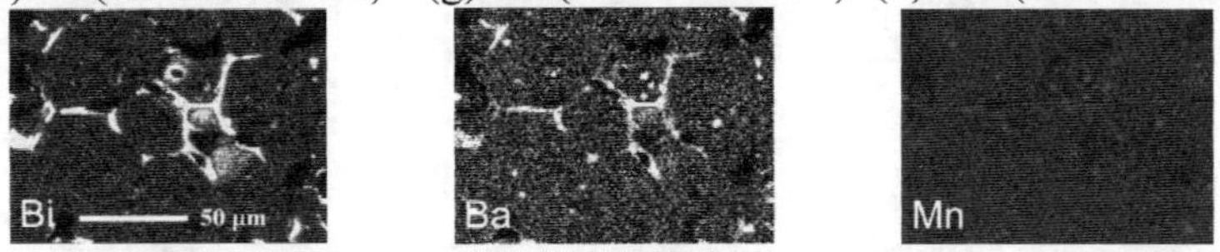

(i) Bi (0.02 mol% Sb) (j) Ba (0.02 mol% Sb) (k) Mn (0.02 mol% Sb)

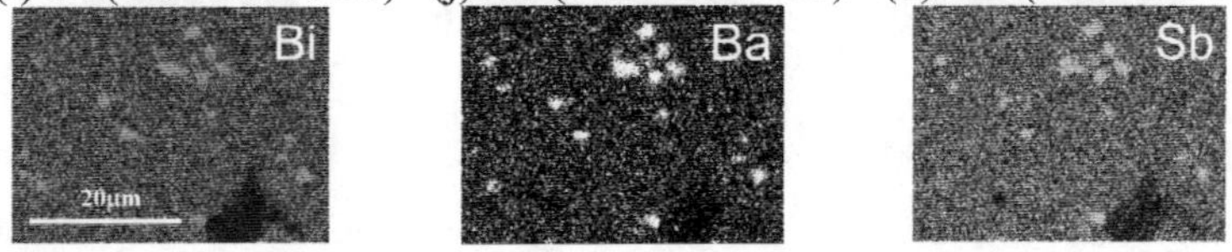

(l) Bi (0.4 mol% Sb) (m) Ba (0.4 mol% Sb) (n) Sb (0.4 mol% Sb)

Figure 5. Elementary mappings and BSE images for various samples.

grain growth is not promoted when Ba and Bi do not coexist in the molten phase during sintering [3]. Therefore, adding a very small amount of Sb (i.e., 0.01, 0.02 mol%) increased the proportion of ZnO grains with diameters over approximately 50 μm, as shown in Figure 1. This is because $Ba_3Mn_2O_8$ was not formed, as shown in Figures 5(g), 5(h), 5(j), and 5(k), and the effective amount of Ba on the ZnO grain growth increased. Furthermore, comparison of Figure 3 and elementary mappings of Mn reveals that the effect of the amount of Ba on ZnO grain growth for 0.02-mol%-Sb-added samples increases compared with the 0.01-mol%-Sb-added samples. On the other hand, both large (i.e., approximately 30 μm) and small (i.e., approximately 10 to 30 μm) ZnO particles appeared for the sample to which 0.1 mol% Sb or more was added, because the addition of a large amount of Sb inhibits ZnO grain growth, as shown in Figures 5(l) through 5(n), due to the formation of compounds or mixtures of Ba, Bi, and Sb.

Figure 6 shows the values of the nonlinearity indexes α of samples to which Sb was not added and samples to which 0.01 mol% Sb and 0.02 mol% Sb were added. As shown in Figure 3, as a result of the addition of Sb, the value of α before electrical degradation increased from approximately 7 to 33 to 40 for the sample having the maximum average grain size obtained by adding Bi. Moreover, the value of α after electrical degradation was improved from approximately 1 for the sample to which Sb was not added to 5 for the 0.02-mol%-Sb-added sample with approximately 0.05 mol% Bi, at which the varistor voltage was low, as discussed later herein. Therefore, the resistance to the electrical degradation for the sample having the maximum average grain size is markedly improved by adding a small amount of Sb, such as 0.02 mol%. One factor in the improvement is the uniformity of the grain size of ZnO obtained by adding Sb. Mn distributed uniformly in ZnO grain boundary has been reported to promote resistance to electrical degradation [6]. Therefore, it is thought that a large amount of Mn is distributed in the grain boundaries because $Ba_3Mn_2O_8$ was not formed by adding Sb [4].

Figure 7 shows the varistor voltages of samples to which Sb was not added and 0.01-mol%-Sb- and 0.02-mol%-Sb-added samples. The varistor voltages for the 0-mol%-Sb- and 0.4-mol%-Bi-added samples were both approximately 12 V/mm. On the other hand, the varistor voltage became low for the sample in which the average grain size became maximal. Value of the varistor voltage for the sample added 0.01 mol% Sb and 0.63 mol% Bi was approximately 33 V/mm and that added 0.02 mol% Sb and 0.5 mol% Bi was approximately 31 V/mm. Although the average grain size of ZnO for the Sb-added sample was larger than that for the sample to which Sb was not added, the varistors voltage became higher. On the other hand, the varistor voltage was not proportional to the reciprocal of the average grain size of ZnO, except for the 0.01-mol%-Sb-added samples. The low varistor voltage for the sample to which Sb was not added is due to the nonuniformity of current flow in the sample. Despite having observed the abnormal growth of ZnO grain for the 0.02-mol%-Sb-added

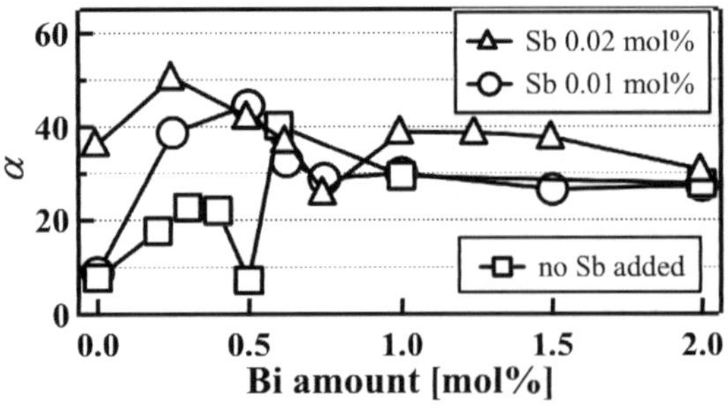

(a) Before electrical degradation

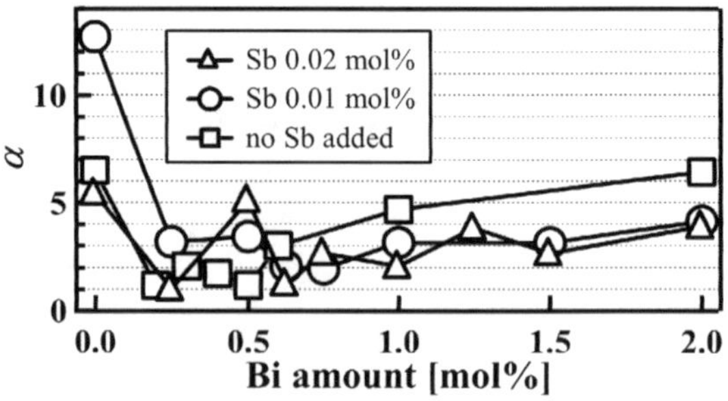

(b) After electrical degradation

Figure 6. Values of α before and after electrical degradation for samples to which no Sb, 0.01 mol% Sb, and 0.02 mol% Sb have been added for various amounts of Bi.

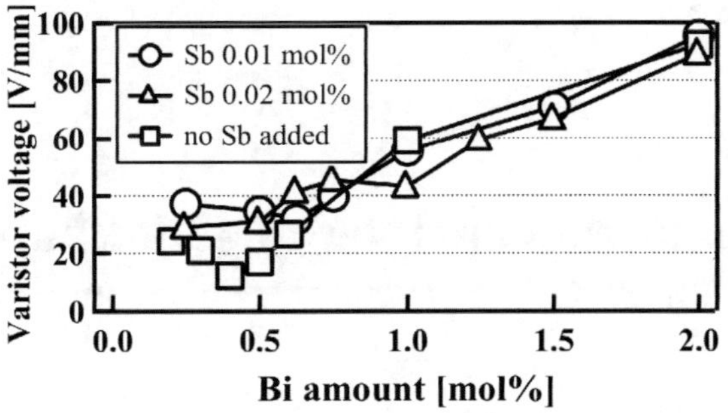

Figure 7. Values of varistor voltages for samples to which no Sb, 0.01 mol% Sb, and 0.02 mol% Sb have been added for various amounts of Bi.

sample, as shown in Figure 1, the varistor voltage was higher than that for the sample to which Sb was not added, and the resistance to the electrical degradation was improved.

4. Summary

The effects of Sb addition to Ba-Co-Mn-added ZnO varistor on ZnO grain growth, varistor voltage, electrical characteristics, and resistance to electrical degradation were investigated changing the amount of addition of Bi. The grain size of ZnO became maximal for the addition of approximately 0.63 mol% and 0.5 mol% Bi for 0.01-mol%-Sb- and 0.02-mol%-Sb-added samples, respectively. The addition of 0.5 mol% Bi and 0.02 mol% Sb provided a low varistor voltage and the highest resistance to electrical degradation, because compounds containing both Ba and Mn, such as $Ba_3Mn_2O_8$, do not form at grain boundaries between ZnO grains by the addition of Sb.

References

[1] Matsuoka M 1971 *Jpn. J. Appl. Phys.* 10 763
[2] Özkan Toplan H and Karakas Y 2002 *Ceramics International* 28 911
[3] Fukumori A, Akiyama Y, Takada M, Sato Y and Yoshikado S 2010 *Key Eng. Maters.* 445 241
[4] Takada M and Yoshikado S 2006 *Key Eng. Maters.* 320 1171
[5] Fukumori A, Akiyama Y, Takada M, Sato Y and Yoshikado S *Key Eng. Maters.* In press
[6] Tanigawa A, Hirota M, Sakaguchi T and Yoshikado S 1998 *T. IEE Japan*, 118-A 169

Sintering and mechanical properties of gadolinium-doped ceria ceramics

K Yasuda*, K Uemura and T Shiota

Ookayama 2-12-1-S7-14, Meguro-ku, Tokyo 152-8552, JAPAN

*E-mail: kyasuda@ceram.titech.ac.jp

Received 22 July 2011, Revised 21 November 2011

Abstract. Gadolinium-doped ceria (GDC) ceramics were made by sintering at various temperatures from 1000˚C to 1400˚C in air. The true density and apparent density were measured to calculate the relative density of GDC ceramics. The change in relative density revealed that densification of GDC ceramics increased up to 1200˚C, and thereafter turned downward. It was suggested that pores were formed at 1300˚C and 1400˚C due to non-stoichiometry of ceria. JIS-type specimens were cut from the sintered body and tested by 4-point bending. Young's modulus and bending strength decreased with increasing the sintering temperature from 1200˚C to 1400˚C, corresponding to the change in the relative density.

1. Introduction

Solid oxide fuel cell (SOFC) is one of the most promising power generators for high efficiency and multi-fuel compatibility[1]; however, its working temperature is as high as 1000ºC because yttria-stabilized zirconia is used as electrolyte. To attain long-term durability of this system, the lower working temperature is needed for next-generation SOFC, and 10-20mol% gadolinium-doped ceria (GDC) ceramics is considered to be one of new electrolytes[2,3].

GDC ceramics has good electrochemical properties even at 800ºC; however, its sintering behavior and mechanical properties have not been fully investigated. Sammes et al.[4] made 20 mol% Gd-doped ceria ceramics by sintering at 1600˚C for 16 hours and obtained 96 % of relative density and 140 MPa of bending strength. Sameshima et al.[5] reported that 20 mol% Sm-doped ceria ceramics (sintered at 1600ºC) had the relative density from 86 to 98 % and fracture strength around 70 MPa. Thus, we do not have enough information on sintering behavior of GDC ceramics and also its fracture strength values are not consistent in the literatures.

In this paper, GDC ceramics were prepared by sintering $Ce_{0.9}Gd_{0.1}O_2$ powder compacts at 1000, 1100, 1200, 1300, 1400ºC, respectively. The true density and apparent density were measured to calculate the relative density of GDC ceramics to make clear its sintering behavior. In addition to it, the bending specimens ($4\times3\times36$ mm^3) were cut from the sintered body, and 4-point bending tests were carried out to discuss the relationship between relative density and mechanical properties of the GDC ceramics.

2. Experimental procedure

2.1. Sample preparation and density measurement

10mol% gadolinium-doped ceria powder was purchased from Anan Kasei Co. Ltd. because this powder is used as de facto standard of GDC powder. The GDC powder (grain size 0.46μm) was compressed in uniaxial pressing at 7 MPa for 2 min, and then compressed again by cold isostatic pressing at 100 MPa for 2 min. The compacts were sintered in air at 1000, 1100, 1200, 1300, 1400°C, respectively. The holding time was 120 min, and heating and cooling rates were 5 °C/min.

According to JIS standards[6,7], rectangular-sectioned specimens were made by cutting from GDC ceramics, and followed by mirror polishing with diamond slurry. The dimensions of the specimens were 4 mm in breadth×3 mm in thickness×36 mm in length. The apparent density was measured by Archimedes method. A small piece of sintered body was ground, and its true density was measured by pycnometry.

2.2. Bending test

4-point bending tests[6,7] were conduced with universal materials testing machine (Shimadzu, Autograph DSC-R-10TS). The upper and lower spans were 10 mm and 30 mm, respectively. The crosshead speed was set to be 0.5 mm/min. Young's modulus E and fracture strength σ_f were calculated from the equations[6,7] below,

$$E = \frac{\Delta W(L-l)^2(2L+l)}{4bt^3\Delta u}, \tag{1}$$

$$\sigma_f = \frac{3W_f(L-l)}{2bt^2} \tag{2}$$

, where L is lower span, l is upper span, b is breadth of the specimen, t is thickness of the specimen, Δu is the net displacement of the specimen after correction by displacement of loading train, ΔW is the load change against Δu, W_f is the fracture load. The number of specimens was around 17 for each sintering condition.

3. Results and discussion

3.1. Sintering behaviour

The sintered body at 1000°C and 1100°C were still porous and weak, so that their bending specimens were not obtained due to breaking down during machining. However, true density and apparent density of these sintered bodies were measured.

Figure 1 and 2 show the true density and apparent density of GDC ceramics. The true density slightly decreases in the specimens sintered at 1300°C and 1400°C. This implies that oxygen vacancy increases from 1300°C. The apparent density also decreases in the specimens sintered at 1300°C and

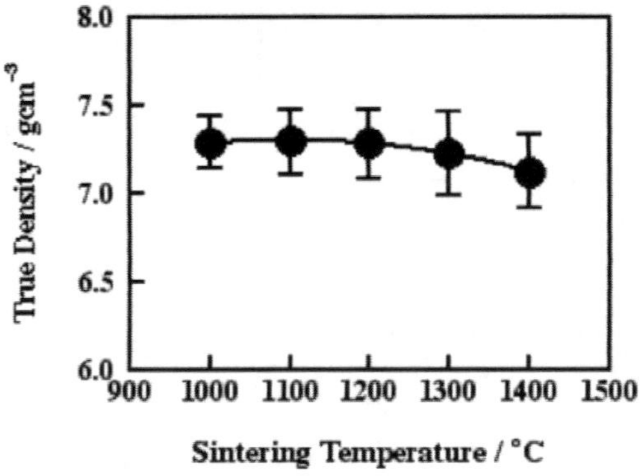

Figure 1. True density of GDC ceramics.

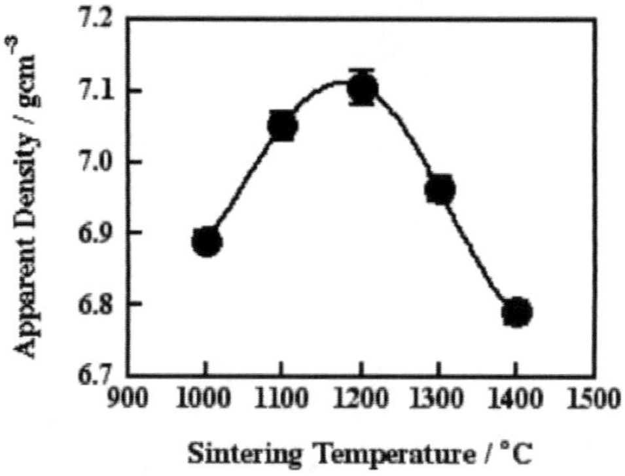

Figure 2. Apparent density of GDC ceramics.

1400°C. This means that pores are actually generated in these specimens.

Figure 3 shows the relative density defined as the ratio of apparent density divided by true density. Densification of GDC ceramics increases with increasing the sintering temperature, however, it turns downward at 1300°C and 1400°C. This means pore formation at high temperatures.

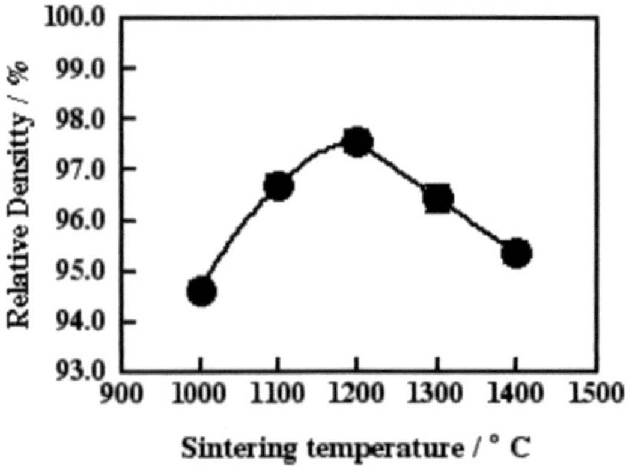

Figure 3. Relative density of GDC ceramics.

Since the authors were anxious about that volatile matter evaporated from the GDC powder at high temperatures, differential thermo gravimetric analysis (TG/DTA) was carried out. As shown in Figure 4, TG curve does not change from 200°C to 1400°C. Its mass loss is less than 1 %, so that we cannot explain the change in relative density (around 3% from 1200°C to 1400°C) by using Figure 4. In addition to this, DTA curve is almost the same as that of a blank test.

As stated in section 3.1 the authors consider that this pore formation is caused by non-stoichiometric phases of ceria at high temperatures[8,9]. The following defect reaction (in Kröger-Vink notation) was reported in these literatures,

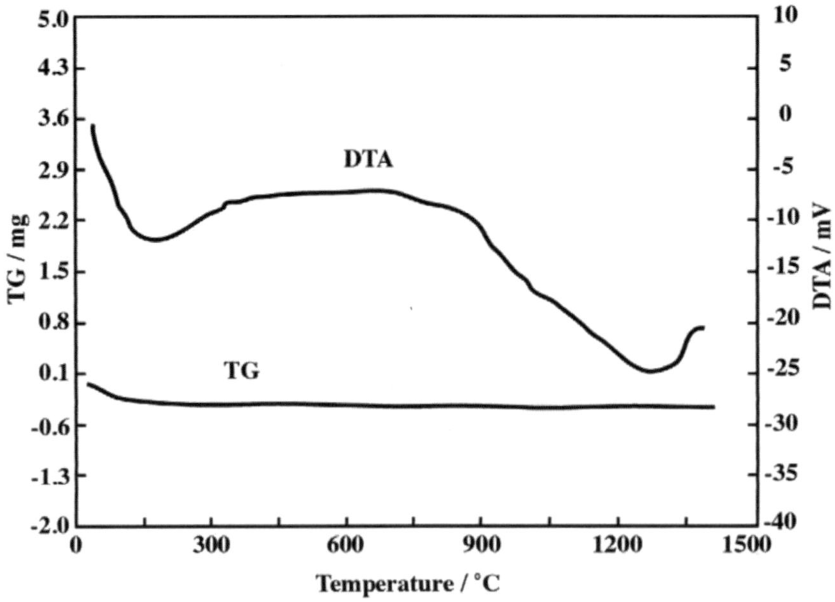

Figure 4. TG/DTA analysis of the GDC powder.

$$2Ce_{Ce}^{\times} + O_O^{\times} \rightarrow 2Ce_{Ce}' + V_O^{\bullet\bullet} + \frac{1}{2}O_2 \qquad (3)$$

The authors have not had any direct evidence of this reaction yet, however, the reaction coincides well with the changes in the true density (viz. oxygen vacancy formation) and also in the relative density (viz. pore formation) at 1300°C and 1400°C.

3.2. Mechanical properties

Figure 5 and 6 show Young's modulus and fracture strength of the GDC ceramics. The bending tests were conducted for the specimens sintered at 1200, 1300, 1400°C only. Young's modulus and fracture strength slightly decrease with increasing the sintering temperature from 1200°C to 1400°C. This tendency can be explained by pore formation at high temperatures, corresponding to the change in the relative density.

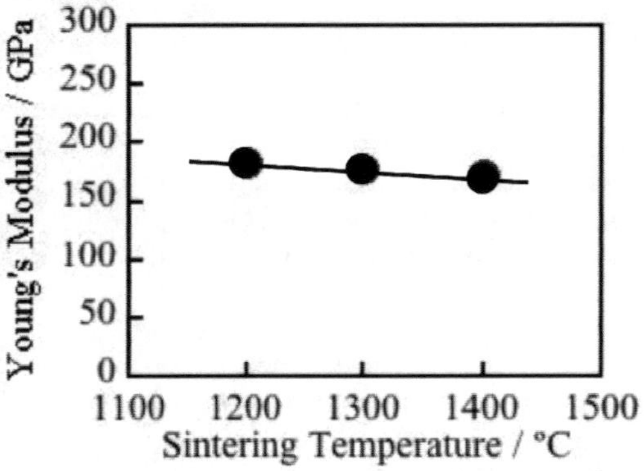

Figure 5. Young's modulus of the GDC ceramics.

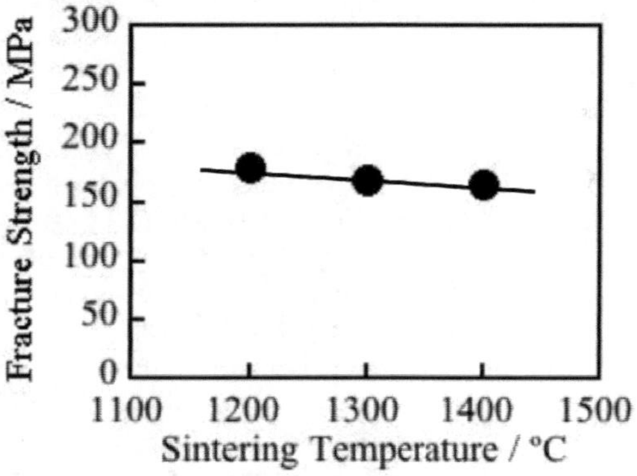

Figure 6. Fracture strength of the GDC ceramics.

4. Conclusion

Densification of GDC ceramics is advanced with increasing the sintering temperature, however, pore formation begins above 1300°C because non-stoichiometric phases of ceria is generated at these temperatures to release an oxygen gas. Therefore, dense GDC ceramics can be obtained by sintering at 1200°C. Young's modulus and fracture strength slightly decrease with increasing the sintering temperature from 1200°C to 1400°C. This corresponds to the change in the relative density.

References

[1] Steele B C H and Heinzel A 2001 *Nature* **414** 345e52
[2] Leng Y J, Chan S H, Jiang S P and Khor K A 2004 *Solid State Ionics* **170** 9
[3] Laukaitis G, Dudonis J and Milcius D 2007 *Materials Science* (*Mendziagotyra*) **13**, 23
[4] Sammes N, Tomsett G and Zhang Y 1996 *Denki Kagaku*, **64** 674
[5] Sameshima S, Ichikawa T, Kawaminami M and Hirata Y 1999 *Materials Chemistry and Physics* **61** 31
[6] JIS R1601
[7] JIS R1602
[8] Greenwood N N 1968 *Ionic Crystals, Lattice Defects and Nonstoichiometry* (London : Butterworth & Co Ltd.) Chapter 5
[9] Bevan D J M and Kordis J 1964 *J.Inorg. Nucl. Chem.* **26** 1509

Effects of Sb, Zr, and Y addition on the electrical characteristics of Bi-based ZnO varistors

T Watanabe, Y Tokoro, Y Sato and S Yoshikado*
Department of Electronics, Doshisha University, Kyoto 610-0321, Japan

*E-mail:syoshika@mail.doshisha.ac.jp

Received 11 August 2011, Revised 24 November 2011

Abstract. The effects of adding Sb, Zr, and Y to a Bi–Mn–Co–Si–Cr–Ni-added ZnO varistor (with the same composition as a commercial varistor) on the varistor voltage, leakage current, and resistance to electrical degradation were investigated. Both the varistor voltage and the nonlinearity index of the voltage–current characteristics α before electrical degradation for samples added 1.0 mol% ZrO_2 and 1.0 mol% Y_2O_3 were maximized at approximately 1.0 mol% Sb_2O_3 and decreased rapidly when the amount of Sb_2O_3 exceeded more than about 1.0 mol%. Samples added 1.0 mol% ZrO_2, and 1.0 mol% Y_2O_3, and less than 0.75 mol% Sb_2O_3 had a comparable leakage current to samples added neither ZrO_2 nor Y_2O_3. Sb is an important element that is necessary to obtain a high varistor voltage, a low leakage current, and an excellent resistance to electrical degradation. However, the same varistor voltage can be obtained by adding Y and reducing the amount of Sb_2O_3 added.

1. Introduction

Excellent resistance to electrical degradation and a high breakdown voltage (varistor voltage) are critical factors for using ZnO varistors in surge protection devices for high-voltage power delivery. The resistance to electrical degradation can be enhanced by adding suitable additives or by thermal annealing [1–3]. If high varistor voltages can be achieved, devices and their cases can be miniaturized, reducing both the amount of resources used and the manufacturing costs. Some commercial ZnO varistors contain Sb, which increases the varistor voltage by reducing the ZnO grain size. However, despite its excellent characteristics, the toxicity of Sb makes it desirable to minimize its use in ZnO varistors. The addition of Y_2O_3 to a Bi-based Sb-added ZnO varistor reduces the average ZnO grain size enabling the varistor voltage to be doubled, but it reduces the resistance to electrical degradation and increases the leakage current [4,5]. The reduction in the resistance to electrical degradation on adding Y to Bi–Mn–Co–Sb–Si–Cr–Ni-added ZnO varistors (with the same composition as a commercial varistor) is due to an increase in the number of willemite (Zn_2SiO_4) particles or to a reduction in the number of spinel $Zn_{2.33}Sb_{0.67}O_4$ particles at the grain boundary between two ZnO grains. The resistance to electrical degradation was improved on adding ZrO_2 [6,7]. Simultaneous addition of Y_2O_3 and ZrO_2 increases the varistor voltage without deteriorating the resistance to electrical degradation or increasing the leakage current [8]. The present study investigates the effect of adding Sb to Bi–Mn–Co–Si–Cr–Ni–Y–Zr-added ZnO varistors with the aim of reducing the amount of Sb added without degrading the electrical characteristics.

Journal of Physics: Conference Series **339** (2012) 012007 doi:10.1088/1742-6596/339/1/012007

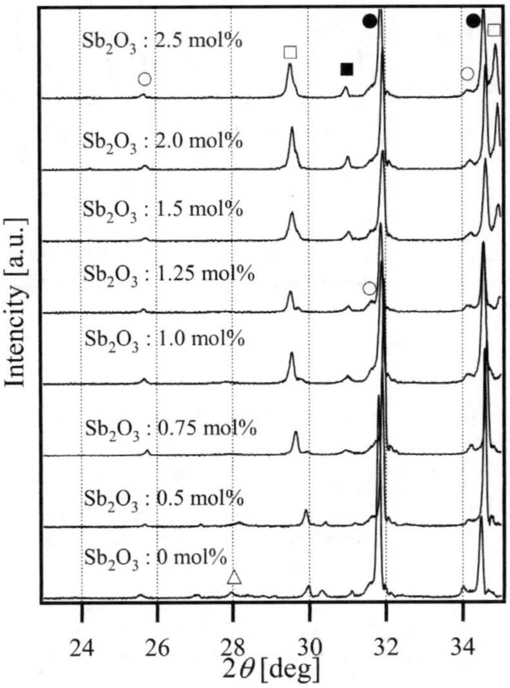

Figure 1. XRD patterns of samples added 1.0 mol% ZrO_2, 1.0 mol% Y_2O_3, and 0−2.5 mol% Sb_2O_3.
△: Bi_2O_3; ●: ZnO; ○: willemite particle
□: spinel particle; ■: unknown

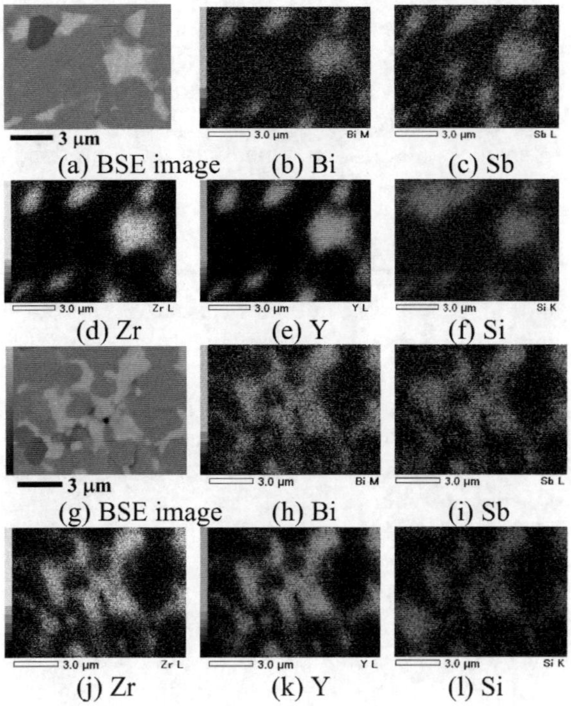

Figure 2. EDS elemental maps for samples added 1.0 mol% ZrO_2, 1.0 mol% Y_2O_3, and (a)−(f) 1 mol% Sb_2O_3 and (g)−(l) 1.5 mol% Sb_2O_3.

2. Experiments

ZnO varistors were fabricated using ZnO (99.9999 purity, MSA), Bi_2O_3 (0.6, 1.2 mol%), MnO_2 (0.55 mol%), Co_3O_4 (0.63 mol%), Cr_2O_3 (0.35 mol%), SiO_2 (2.65 mol%), and NiO (0.6 mol%) powders as basic additives and Sb_2O_3 (0−2.5 mol%), ZrO_2 (0, 1 mol%), and Y_2O_3 (0−2.5 mol%) powders. Powder mixtures were wet mixed for 24 h with ethanol. A pressed tablet of the mixture was then calcined at 600°C for 5 h in air and then ground in an agate mortar. The ground powder was pressed into a 20-mm-diameter tablet at a pressure of 320 MPa. The tablets were sintered at 1150°C for 3 h in air and then allowed to cool to room temperature. To fabricate samples for voltage−current (V−I) characteristics measurements, aluminum electrodes were formed on the two end faces of each sample (3×3 mm^2) by vacuum evaporation. The V−I characteristics were measured in a current density range of approximately 3×10^{-7}–3×10^{-3} A/cm^2 by applying a constant DC current. The nonlinearity index α of the V−I characteristics was estimated using the equation $V = V_N(I/I_N)^{1/\alpha}$, where V and V_N are normalized voltages per 1 mm and I and I_N are normalized currents per 1 cm^2. V_N is the voltage for $I = I_N$; V_N for $I_N = 1\times10^{-3}$ A/cm^2 is the so-called varistor voltage. The nonlinearity of the V−I characteristics is high, giving rise to a high value of α. The value of α was calculated from the V−I characteristics in the current density range of 1×10^{-4}–1×10^{-3} A/m^2. The uniformity of the current inside a sample was estimated from the form of the V−I characteristics. When the current density varies inside a sample, the V−I characteristics may vary erratically in the breakdown region. Therefore, the reproducibility of the V−I characteristics was confirmed by measuring 5 to 10 samples. Moreover, when voltages lower than the varistor voltage are applied to a sample, the V−I characteristics correspond to the pre-breakdown region and the V−I characteristics are linear. In this study, the leakage current is defined as the current when 80% of the varistor voltage was applied to a sample. The electrical degradation was achieved by applying a DC current of 100 mA/cm^2 in air for 30 min.

The elemental compositions of the samples were analyzed by scanning electron microscopy and energy-dispersive X-ray spectroscopy (EDS) and by using a backscattered electron (BSE) detector. The crystal structure was analyzed by X-ray diffraction (XRD; Cu $K_{\alpha 1}$). The relative integral intensity *RII* of the diffraction peak of a given compound is defined as the ratio of the sum of the integral intensity of all diffraction peaks of the compound to the integral intensity of the ZnO peak.

3. Results and discussion

Figure 1 shows XRD patterns of samples added 1.0 mol% ZrO_2, 1.0 mol% Y_2O_3, 0–2.5 mol% Sb_2O_3, and 0.6 mol% Bi_2O_3. Diffraction peaks were observed at $2\theta \approx 27.7°$ for δ-Bi_2O_3, at $2\theta \approx 25.6$, 31.2, and 33.8° for willemite particles, at $2\theta \approx 29.4$ and 34.6° for spinel particles, at $2\theta \approx 30.7°$ for an unknown compound that contained significant amounts of Sb, Y, and Zr and a small amount of Bi, and at $2\theta \approx 31.6$ and 34.4° for ZnO [7]. The intensity of the diffraction peak of Bi_2O_3 decreased as Sb_2O_3 was added up to approximately 1.25 mol% and the diffraction peaks disappeared when more than approximately 1.5 mol% Sb_2O_3 was added. The intensities of diffraction peaks for spinel particles and the unknown compound increased as Sb_2O_3 was added up to approximately 1.0 mol% and became constant when more than approximately 1.25 mol% Sb_2O_3 was added. The diffraction peaks for spinel particles shifted to lower diffraction angles when less than approximately 1.0 mol% Sb_2O_3 was added. The intensity of the diffraction peak for willemite particles did not change noticeably on the addition of about 1.0 mol% Sb_2O_3.

Figure 2 shows BSE images and EDS elemental maps for Bi, Sb, Zr, Y, and Si of samples added 1.0 mol% ZrO_2, 1.0 mol% Y_2O_3, 0.6 mol% Bi_2O_3 and ((a)–(f)) Sb_2O_3 0.75 mol% and ((g)–(l)) 1.5 mol% Sb_2O_3. The gray-scale brightness in both the BSE and EDS images increases with increasing amount of an element and with increasing atomic weight of the element. Bi is present at grain boundaries and coexists with segregated particles such as spinel particles, willemite particles, and the unknown compound at the triple points. The EDS elemental map for Bi differs slightly from those for Sb and Si for the sample added 0.75 mol% Sb_2O_3, but it is very similar to those for Sb, Zr, and Y for the sample added 1.5 mol% Sb_2O_3. The XRD and EDS analysis results reveal that Bi is incorporated in spinel particles until δ-Bi_2O_3 eventually disappears as the amount of Sb_2O_3 increased.

Figure 3 shows the varistor voltage, the nonlinearity index before electrical degradation α_{bef}, and the relative integral intensity *RII* of Bi_2O_3 of samples added 0–2.5 mol% Sb_2O_3 and 0.6 mol% Bi_2O_3 and of samples added 0–2.5 mol% Sb_2O_3, 0.6 mol% Bi_2O_3, 1.0 ZrO_2, and 1.0 mol% Y_2O_3. The varistor voltage of samples added neither ZrO_2 nor Y_2O_3 increased with increasing amount of Sb_2O_3. Samples to which only Y_2O_3 had not been added exhibited the same behavior as those added Sb_2O_3

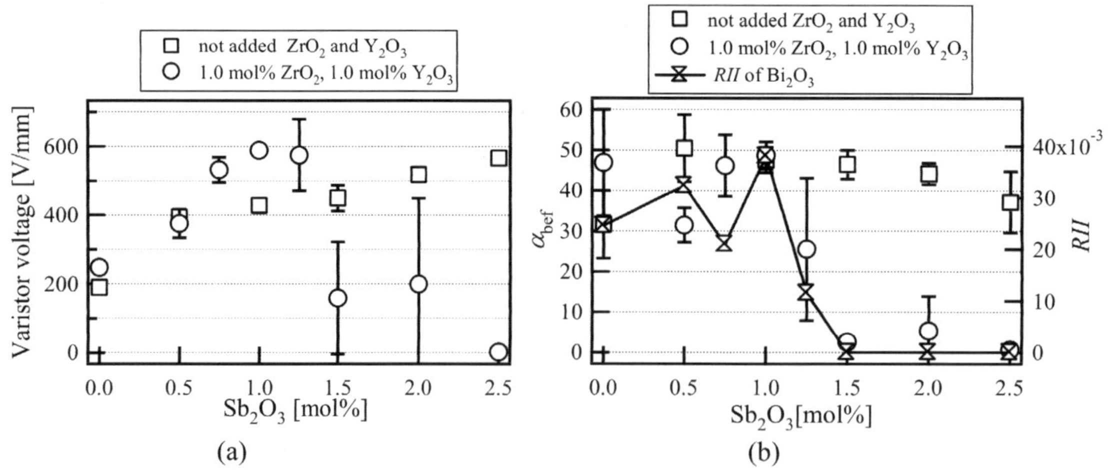

Figure 3. (a) Varistor voltage (b) α_{bef} and *RII* of samples added 0–2.5 mol% Sb_2O_3 and 0.6 mol% Bi_2O_3 and of samples added 0–2.5 mol% Sb_2O_3, 0.6 mol% Bi_2O_3, 1.0 mol% ZrO_2, and Y_2O_3 1.0 mol%.

(data not shown). On the other hand, samples added both ZrO_2 and Y_2O_3 had a maximum varistor voltage of approximately 600 V/mm for 1.0 mol% Sb_2O_3 and their varistor voltage decreased rapidly as the amount of Sb_2O_3 increased above 1.0 mol%. This maximum varistor voltage is almost the same as the asymptotic value of the varistor voltage of the samples added neither ZrO_2 nor Y_2O_3 with increasing amount of Sb_2O_3. Both α_{bef} and *RII* of samples added ZrO_2 and Y_2O_3 showed the same behavior with the addition of Sb_2O_3 and the nonlinearity disappeared when more than approximately 1.5 mol% Sb_2O_3 was added. Since the *V–I* characteristics are nonlinear even when 2.5 mol% Y_2O_3 is added (see Figure 4) when less than approximately 1 mol% Sb_2O_3 is added, the amount of Sb_2O_3 in the presence of Zr and/or Y is considered to be the main cause for the disappearance of the nonlinear *V–I* characteristics. The nonlinear *V–I* characteristics did not disappear for the sample added 1.2 mol% Bi_2O_3 and 1.5 mol% Sb_2O_3. The reduction in both α_{bef} and the amount of δ-Bi_2O_3 with the addition of more than approximately 1.25 mol% Sb_2O_3 demonstrates that Sb inhibits Bi_2O_3 from forming deep interfacial impurity levels at grain boundaries [1].

Figure 5 show the leakage current dependence on the amount of Sb_2O_3 for samples added 0–2.5 mol% Sb_2O_3 and 0.6 mol% Bi_2O_3 and for samples added 0–2.5 mol% Sb_2O_3, 0.6 mol% Bi_2O_3, 1.0 mol% ZrO_2, and Y_2O_3 1.0 mol%. Samples added ZrO_2, Y_2O_3, and less than 0.75 mol% Sb_2O_3 had comparable leakage currents to samples added neither ZrO_2 nor Y_2O_3. The increase in the leakage current for samples added ZrO_2, Y_2O_3, and more than 1.25 mol% Sb_2O_3 is due to the disappearance of the nonlinear *V–I* characteristics.

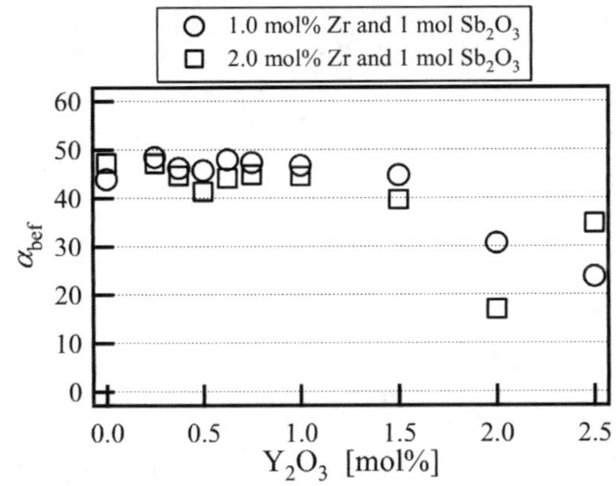

Figure 4. α_{bef} for samples added 1.0 or 2.0 mol% ZrO_2, and 1.0 mol% Sb_2O_3.

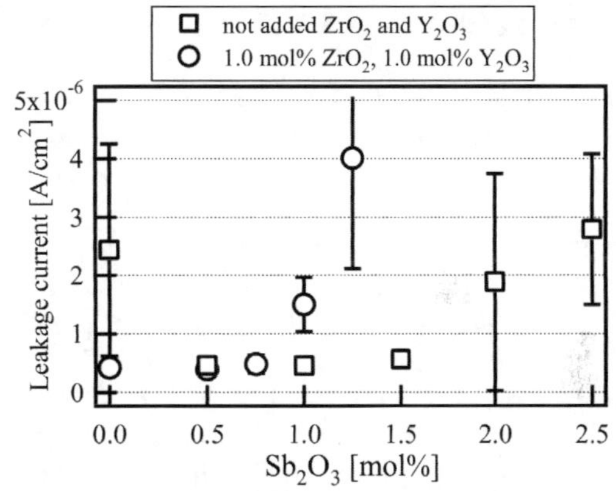

Figure 5. Leakage current for samples added 0–2.5 mol% Sb_2O_3 and 0.6 mol% Bi_2O_3 and samples added 0–2.5 mol% Sb_2O_3, 0.6 mol% Bi_2O_3, 1.0 mol% ZrO_2, and 1.0 mol% Y_2O_3.

Figure 6 shows the nonlinearity index α_{bef} after electrical degradation for samples added 0–2.5 mol% Sb_2O_3 and 0.6 mol% Bi_2O_3 and samples added 0–2.5 mol% Sb_2O_3, 0.6 mol% Bi_2O_3, 1.0 mol% ZrO_2, and 1.0 mol% Y_2O_3. α_{bef} had a maximum value for 1 mol% Sb_2O_3 for all samples.

The sample added the optimal amount of Sb_2O_3 (1 mol% for samples added 0.6 mol% Bi_2O_3) exhibited a high varistor voltage and a high resistance to electrical degradation. However, its leakage current needs to be further reduced.

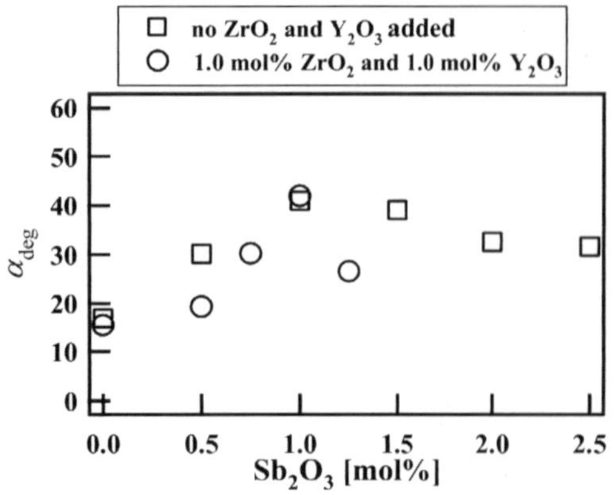

Figure 6. Leakage current for samples added 0−2.5 mol% Sb_2O_3 and 0.6 mol% Bi_2O_3 and for samples added 0−2.5 mol% Sb_2O_3, 0.6 mol% Bi_2O_3, 1.0 mol% ZrO_2, and 1.0 mol% Y_2O_3.

4. Summary

Bi is incorporated in spinel particles until δ-Bi_2O_3 eventually disappears as the amount of Sb_2O_3 increased. The reduction in both the nonlinearity index and the amount of δ-Bi_2O_3 on the addition of more than approximately 1.25 mol% Sb_2O_3 demonstrates that Sb inhibits Bi_2O_3 from forming deep interfacial impurity levels at grain boundaries. The sample added the optimal amount of Sb_2O_3 (1 mol% for samples added 0.6 mol% Bi_2O_3) exhibits a high varistor voltage and a high resistance to electrical degradation. Sb is important for obtaining a high varistor voltage and a high resistance to electrical degradation. However, the same varistor voltage can be obtained by adding ZrO_2 and/or Y_2O_3 and reducing the amount of Sb_2O_3 added.

References

[1] Eda K, Iga A and Matsuoka M 1980 *J. Appl. Phys.* **51** (5) 2678
[2] Kim C-H and Kim J-H 2004 *Ceram. Soc.* **24** 2537
[3] Takada M and Yoshikado S 2009 *J. Euro. Ceram. Soc.* **30** 531
[4] Takemura T, Kobayashi M, Takada Y and Sato K 1987 *J. Am. Ceram. Soc.* **70** 237
[5] Bernik S, Mac S and Ai B 2001 *J. Euro. Ceram. Soc.* **21** 1875
[6] Akiyama Y, Takada M, Fukumori A, Sato Y and Yoshikado S 2010 *Key Eng. Mater.* **421–422** 209
[7] Akiyama Y, Fukumori A, Takada M, Sato Y and Yoshikado S 2010 *Key Eng. Mater.* **445** 237
[8] Watanabe T, Fukumori A, Akiyama Y, Sato Y and Yoshikado S 2011 *Key Eng. Mater.* **485** 257

Synthesis and characterization of poly(sodium acrylate)/ bentonite superabsorbent composite

Yiming Xie, Yuelin Wei, Yunfang Huang, Jianming Lin, and Jihuai Wu*
Eng. Res. Center of Environment-Friendly Functional Mater., Ministry of Education, Inst. of Materials Physical Chemistry, Huaqiao University, Quanzhou, 362021, China

*E-mail: jhwu@hqu.edu.cn

Received 4 August, 2011, Revised 22 November 2011

Abstract. Poly(sodium acrylate)/bentonite superabsorbent composite with water absorbency 1562 $g{\cdot}g^{-1}$ was synthesized by inverse suspension polymerization. The introduction of bentonite improves the water absorbency and facilitates the particle size even distribution of the composite. The network structure in the superabsorbent hydrogel is confirmed.

1. Introduction

Superabsorbent polymer is a special polymeric material with hydrophilic groups and loosely crosslinked networks, it is able to absorb hundreds to thousands times its own mass of water, and the absorbed water is hardly removed under some pressure. Because of its excellent properties, superabsorbent polymer has aroused considerable research and applied in the field of medicine, healthcare, agriculture, artificial intelligence materials, and so on [1-4], since the first superabsorbent polymer was reported by the Department of Agriculture U. S. in 1976 [5].

Clay is a layered aluminium silicate with exchangeable cations in its interlayer and active OH groups on its surface. It is expected that the network structure and hydrophilic groups of the superabsorbents are improved by polymerizing polymer with clay. Consequently, comprehensive absorbent property of the superabsorbent is improved and the production cost is reduced [6-9]. In previous research [10, 11], we reported the synthesis and swelling behaviours of polymer/clay superabsorbent composites prepared by aqueous solution methods. But, due to high water absorbency for superabsorbent composite, the abundant water absorbed in the composite is hardly removed from the products in the production process of superabsorbent composite by aqueous polymerization. In order to overcome the problem, poly(sodium acrylate)/bentonite superabsorbent composite is synthesized by inverse-suspend polymerization method, the superabsorbent composite shows a water absorbency of 1562 $g{\cdot}g^{-1}$.

2. Experimental part

2.1. Materials

Bentonite micropowder (Linan Bentonite Factory, Zhejian, China) milled through 320-mesh screen, then, dried at 105 ºC for 24 h prior to use. Other agents were purchased from Shanghai Chemical Reagent Factory, Shanghai, China. Acrylic acid monomer was purified by freezing before use.

Initiator potassium persulfate, crosslinker N, N'–methylenebis acrylamide, solvent cyclohexane, surfactant span 60 (sorbitan monostearate) and sodium hydroxide were used without further purification. All solutions were prepared by using distilled water as solvent.

2.2. Preparation of poly(sodium acrylate)/bentonite superabsorbent composite

Poly(sodium acrylate)/bentonite superabsorbent composite was prepated by inverse suspension polymerization according to following steps. A sodium acrylate solution was prepared by slowly dropping acrylic acid monomer into a sodium hydroxide solution to achieve a desired neutralization degrees at 0 °C (cooled by ice-water bath) under stirring. A predetermined amount of bentonite micropowder was added to the sodium acrylate solution. Then, solvent cyclohexane and surfactant span 60 were added. Under a nitrogen atmosphere, the crosslinker N, N-methylenebisacrylamide and the initiator potassium persulphate were added to the above mixture solution. The mixed solution was stirred at room temperature for 30 min, and heated slowly to 65 °C with a water bath under stirring. After the polymerization reaction was carried on for 90 min, a jell-like intermediate product was filtered out, washed with methanol three times and dried in an oven at 70 °C more than 24 h to a constant weight, then milled and screened, a powdered poly(sodium acrylate)/bentonite superabsorbent composite product thus was obtained.

2.3. Characterization and measurement

The morphology of sample was observed with an S-3500N SEM instrument (Hitachi, Japan). The water absorbency of the sample was measured by immersing the dried powdered superabsorbent composite of 3 g in distilled water of 500 ml at room temperature for at least 3 h to reach swelling equilibrium, which resulted in the absorption of water inside the superabsorbent composite and the formation of a swollen hydrogel. Unabsorbed water was separated from the swollen sample by filtered over a 100-mesh screen. The water absorbency (Q_{H2O}) of the sample was determined according to the following equation:

$$Q_{H_2O} = \frac{(m_2 - m_0) - (m_1 - m_0)}{(m_1 - m_0)} = \frac{m_2 - m_1}{m_1 - m_0} \qquad (1)$$

Where m_0, m_1 and m_2 are the weights (g) of the clay in the sample, the dry sample and the swollen sample, respectively.

3. Results and Discussion

3.1. Influence of bentonite amount on the water absorbency of the superabsorbent composite

The influence of the amount of bentonite powder in poly(sodium acrylate)/bentonite composite on the water absorbency (Q_{H2O}) is shown in Figure 1. It can be seen that the Q_{H2O} increases from 1330 to 1562 $g \cdot g^{-1}$ when the bentonite percentage in the composite increases from 0 to 20 wt.%. Beyond a bentonite amount of 20 wt. %, the Q_{H2O} decreases. The composite is formed by polymerizing acrylate onto bentonite. Bentonite powder as a network point plays an important role in the formation of the composite. [9, 10, 12, 13] When the amount of bentonite is suitable, such as amounts of 0–20 wt. %, the crosslinking density of the composite does not change or only changes a little, the superabsorbent composite has a enough space to absorb and hold water molecules. However, with the increase of bentonite amount in the composite, such as higher than 20 wt.%, the composite's crosslinking density increases and network space shrinks, which results in the difficult permeation of water molecules into the composite. Thereby, the Q_{H2O} value decreases with the increase of bentonite percentage (higher than 20 wt.%) in the composite.

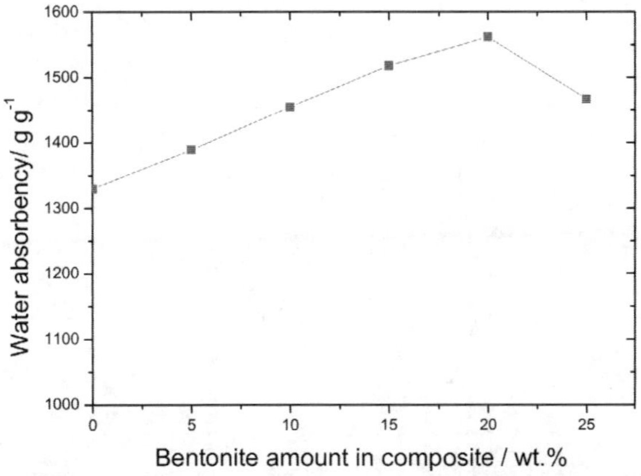

Figure 1. Variation of Q_{H2O} with bentonite amount in poly(acrylate sodium)/bentonite composite. Initiator of 0.2 wt.%, crosslinker of 0.01 wt.%, and polymerization at 60 °C.

3.2. Size and shape of superabsorbent composite

Figure 2 shows the SEM images of poly(sodium acrylate) superabsorbent polymer (a) and poly(sodium acrylate)/bentonite superabsorbent composite (b). The polymer shows [Figure 2(a)] a spherical and ellipsoidal shape with a wide size distribution, the diameter of the small particles is 6-10 μm and the large particles is 100-120 μm. The composite [Figure 2(b)] shows a spherical shape with a narrow size distribution and the diameter in the range of 4-14 μm. The narrower size distribution for the composite is due to that bentonite micropowder acts as a dispersant in the inverse suspension polymerization process. The surface shapes of the particles in Figure 2 (b) are irregular and rough compared with the particles in Figure 2 (a), which maybe result from that bentonite powders are irregular and rough.

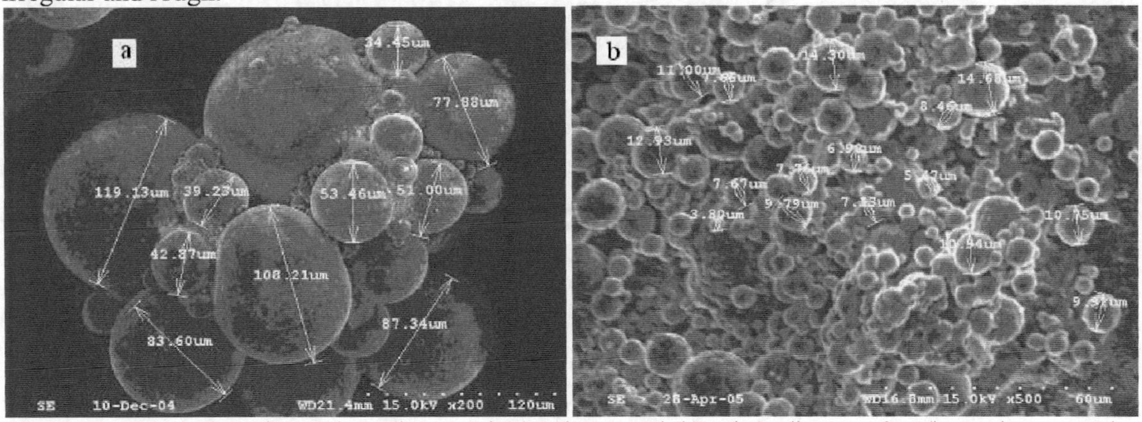

Figure 2. SEM images of (a) poly(sodium acrylate) polymer and (b) poly(sodium acrylate)/bentonite composite

3.3. Microstructure of superabsorbent

The interior morphology of the superabsorbent composite was observed with SEMs. In order to see the interior structure of superabsorbent clearly, two samples were choose. One was the sample just synthesized without drying and swelling, the SEM image is shown in Figure 3 (a); the other was the sample swollen fully at room temperature, keeping at 0 °C, measuring under vacuum at 0 °C, the SEM image is shown in Figure 3 (b). The interior morphology of dried superabsorbent composite is small and irregular pores [Figure 3 (a)]. In contrast, the superabsorbent composites hydrogel presents a macroporous network structure [Figure 3 (b)]. Due to be not swollen fully, the polymer chain of the sample (a) shows a shrinked state and small pores. On the other hand, due to be swollen fully at room temperature, the polymer chains of the sample (b) unfold fully, the space of network is filled with water. Then, the sample (b) is kept at 0 °C, the water is frozen inside of the polymer network and can not evaporate. Under vacuum and 0 °C measuring condition, the water in the polymer network evaporate but the swollen hydrogel's space structure is still kept, superabsorbent hydrogel shows its real structure, which presents a macroporous network structure. From Figure 3, the network structure of the superabsorbent hydrogel is confirmed.

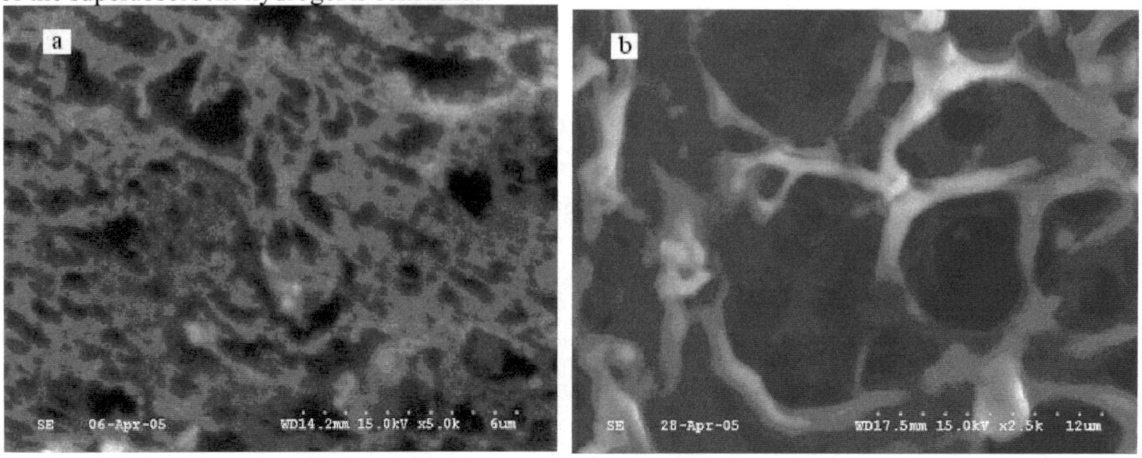

Figure 3. SEM images of dried superabsorbent composite (a) and fully swollen superabsorbent composite hydrogel (b).

4. Conclusion

In summary, poly(sodium acrylate)/bentonite superabsorbent composite with water absorbency of 1562 $g \cdot g^{-1}$ was synthesized by inverse suspension polymerization. The introduction of bentonite improves the water absorbency and facilitates the particle size even distribution of the composite. The network structure in the superabsorbent hydrogel is confirmed.

Acknowledgement

The authors thank for the joint support by the National High Technology Research and Development Program of China (No. 2009AA03Z217), and the National Natural Science Foundation of China (No. 90922028, 50842027).

References

[1] Omidian H, Roccaa J G and Park K 2005 *J. Controlled Release* **102** 3
[2] Mohan Y M, Murthy P S K and Raju K M 2005 *React. Funct. Polym.* **63** 11
[3] Pourjavadi A, Ghasemzadeh H and Soleyman R 2007*J. App. Polym. Sci.* **105** 2631
[4] Sahoo P K, Sahu G C, Rana P K and Das A K 2005 *Adv. Polym. Tech*, **24** 208
[5] US 3 981 100 (1976), United States Dept. of Agriculture, invs.: Weaver M O, Bagley E B, Fanta G F and Doane W M *Chem. Abstr.* **1976**, 85, 161504n
[6] Wu J, Lin J and Wei C 2000 *Macromol. Rapid Commun.* **21** 1032
[7] Wu J, Wei Y and Lin J 2003 *Polymer* **44** 6513
[8] Pourjavadi A, Barzegar S and Zeidabadi F, 2007 *React. Funct. Polym.* **67** 644
[9] Yoshimura T, Yoshimura R, Seki C and Fujioka R 2006 *Carbohydrate Polymers* **64** 345
[10] Lin J, Wu J and Yang Z 2001 *Macromol. Rapid Commun.* **224** 22
[11] Wu J, Lin J and Li G 2001 *Polym. Int.* **50** 1050
[12] Haraguchi K, Takehisa T and Fan S 2002 *Macromolecules* **35** 10162
[13] Haraguchi K, Farnworth R and Takehisa T 2003 *Macromolecules* **36** 5732

Effect of pH on formation of γ-MnO_2 from Zn-bearing $MnSO_4$ solution

C X Liu, L Z Zhang, L Xiang*
Department of Chemical Engineering, Tsinghua University, Beijing, 100084, China

*E-mail: xianglan@mail.tsinghua.edu.cn

Received 14 August 2011, Revised 28 November 2011

Abstract. Manganese dioxide has attracted much attention owing to its distinctive properties. A facile method was developed to synthesis γ-MnO_2 by the reaction of $Na_2S_2O_8$ with Zn-bearing $MnSO_4$, keeping the molar ratio of Mn to S as 1:2-4. The oxidation of Mn^{2+} increased from 62.3% to 99.9% with the increase of the final solution pH from 0.55 to 4.45. The needle-like γ-MnO_2 particles with a length of 200-500 nm and a diameter of 10-30 nm were synthesized in the initial solution pH range of 3.5-4.0. γ-MnO_2 particles with poor crystallinity and porous morphology were synthesized in the initial solution pH range of 6.0-6.5. The Zn^{2+} adsorption capacities of the γ-MnO_2 formed at initial pH=3.5 and pH=6.0 were 131.8 $mg.g^{-1}$ and 527.4 $mg.g^{-1}$, respectively.

1. Introduction

Manganese dioxides have attracted great attention owing to their special physical and chemical properties and wide application in the fields as magnetic, energy, redox catalytic and battery materials [1-2]. α-, β-, γ- and δ- MnO_2 polymorphs existed with the varying linkage of [MnO_6] octahedral units [3]. Wang [4] showed that the order of the capacitance values of MnO_2 was β-MnO_2 (57.7 $F\cdot g^{-1}$) <α-MnO_2 (103.9 $F.g^{-1}$) <γ-MnO_2 (237.6 $F.g^{-1}$) and γ-MnO_2 was more likely to have favourable capacitive behaviour. Zhang [5] found that compared with crystalline γ-MnO_2, the amorphous γ-MnO_2 exhibited better capacitance properties. Huang [6] reported that γ-MnO_2 can improve the specific capacitance of some active material owing to the rich hydroxyl band on the surface, which acted as the conductors for the proton transfer.

γ-MnO_2 with varying morphologies as wires, rods, belts and tubes have been synthesized [7-11]. Zhao [12] prepared hollow micro-sphere γ-MnO_2 with diameter of 2-3μm via the decomposition of $MnCO_3$. Chen [13] etc. synthesized the γ-MnO_2 nano-tubes by decomposition of $Mn(OH)_2$ in the anodic aluminium oxide(AAO) templates and thermally annealing at 500°C. γ-MnO_2 was also synthesized by hydrothermal treatment at 120°C for 2h [14].

In this work, a facile method was developed to synthesize γ-MnO_2 via the oxidation of Mn^{2+} from Zn-bearing solution. The influence of the initial and final solution pH on the oxidation of Mn^{2+} and morphology of γ-MnO_2 were discussed.γ-MnO_2 with different crystallinites and adsorption capacities of Zn^{2+} were prepared at varying solution pH.

2. Experimental

2.1. Synthesis of γ-MnO_2

In a typical experiment, certain amount of $Na_2S_2O_8$(s) was added to 0.1-0.5 mol.L^{-1} $MnSO_4$ containing 0.5-2.0 mol.L^{-1} $ZnSO_4$, keeping the molar rate of Mn^{2+} and $Na_2S_2O_8$ as 1:2-4. Certain amount of H_2SO_4 or NaOH was used to change the initial pH of the solution. The mixtures were stirred at 90℃ for 2.0 h, and then another certain amount of H_2SO_4 or NaOH was added before 1.0 h more reaction, in order to change the final pH of the solution as wanted. And then the precipitates were filtrated, washed, and collected for characterization.

2.2. Analysis

The morphologies of the samples were examined with the field emission scanning electron microscopy (JSM7401F, JEDL, Japan). The structure and composition of the sample were identified by the X-ray powder diffractometer (D8 advance, Brucker, Germemy). The content of Zn^{2+} was analyzed by EDTA titration method, and the solution pH was detected by pH meter (DELTA 320, METTLER-TOLEDO, China). The initial concentration of Zn^{2+} and volume of the solution were c0 and v0, while the final concentration of Zn^{2+} and volume of the filtrate after the filtration were c1 and v1. Thus the adsorption rate of Zn^{2+} could be calculated as follows:

$$Adsorp_{Zn} = (c0*v0-c1*v1)/(c0*v0).$$

3. Results and discussions

3.1. Oxidation of Mn^{2+}

Mn^{2+} was oxidized to MnO_2 by $Na_2S_2O_8$:

$$MnSO_4 + Na_2S_2O_8 + 2H_2O = MnO_2 \downarrow + Na_2SO_4 + 2H_2SO_4$$

Figure 1 shows the variations of the solution pH and the oxidation of Mn^{2+} with the reaction time at 90°C. With the increase of the reaction time from 0 to 3.5 h, the solution pH decreased from 3.42 to 0.55, and the oxidation of Mn^{2+} increased from 0 to 62.3%.

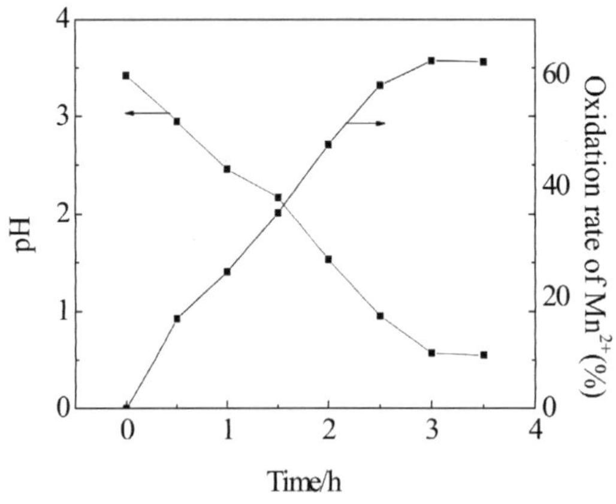

Figure 1. Variation of the solution pH and Mn^{2+} oxidation with reaction time

Figure 2 shows the influence of the final solution pH and temperature on the oxidation of Mn^{2+}. After action at 90°C for 3h, the final pH increased from 2.0 to 4.5 and the oxidation of Mn^{2+} from 72.5% to 99.9%. At the final solution pH 4.0, the oxidation of Mn^{2+} increased from 34% to 99.9% as the increase of the temperature from 50°C to 90°C. Higher temperature and higher final solution pH favored the oxidation of Mn^{2+} to MnO_2. Table 1 shows the influences of the initial and final solution pH on the oxidation of Mn^{2+}. In the case of the final solution pH 4.0-4.1, more than 99.9% of Mn^{2+} was oxidized quickly to MnO_2 despite of the change of the initial solution pH from 1.67 to 6.14; In the case of the initial solution pH 3.4, the oxidation of Mn^{2+} increased from 62.3% to 99.9% as the final solution pH increased from 0.55 to 4.45. The above results showed that the oxidation of Mn^{2+} connected closely with the final solution pH and was independent of the initial solution pH.

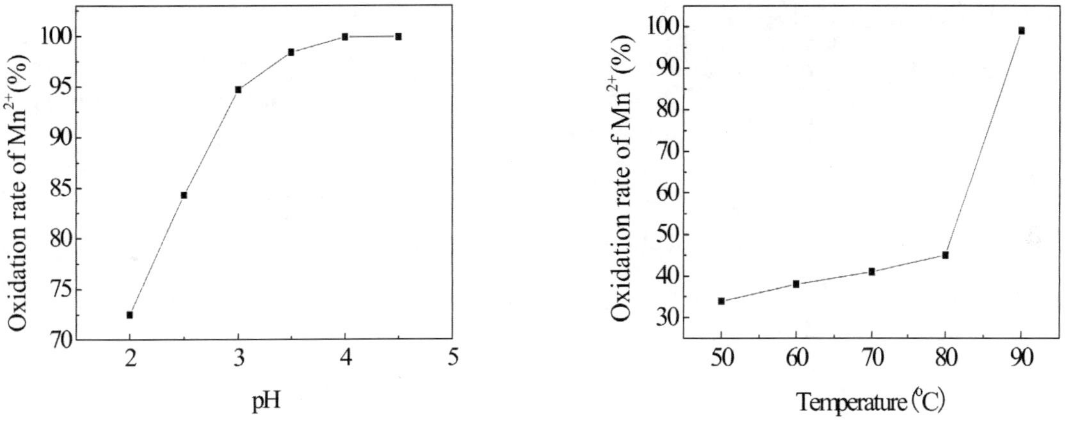

Figure 2. Influences of the final solution pH and temperature on oxidation of Mn^{2+}

Table 1. Variation of oxidation of Mn^{2+} with the initial and final solution pH

Initial pH	Final pH	Oxidation of Mn^{2+}/%	Initial pH	Final pH	Oxidation of Mn^{2+} /%
6.14	4.10	99.9	3.42	0.55	62.3
4.50	4.03	99.9	3.40	2.07	72.6
3.42	4.07	99.9	3.41	3.47	98.4
1.67	4.05	99.9	3.42	4.45	99.9

3.2 Morphologies and XRD patterns of γ-MnO_2

Figure 3 shows the influence of the initial solution pH on the morphology and XRD patterns of γ-MnO_2. In the case of the initial solution pH 3.5-4.0, the needle-like γ-MnO_2 particles with a diameter of 10-30 nm and a length of 200-500 nm were formed after reaction at 90°C for 3 h. In the case of the initial pH 6.0-6.5, porous and poor crystalline γ-MnO_2 formed. The weak and wide diffraction peaks emerged at 2θ = 37.3°, 42.2°, and 56.5° were ascribed to (020), (121), and (221) planes of γ-MnO_2 , respectively.

The solution pH was increased by NaOH at varying reaction time to reveal the influence of pH on the formation of γ-MnO_2 and the results were shown in Figure 4. The experimental results indicate that with the step increase of solution pH after 2 hours of reaction, the crystallinity of γ-MnO_2 increased. Keeping the final pH at around 4.0, 99.9% of Mn^{2+} was converted to MnO_2 in all of the cases. The above work shows that the crystallinity of γ-MnO_2 was connected with the initial pH solution and the oxidation of Mn^{2+} was connected with the final solution pH.

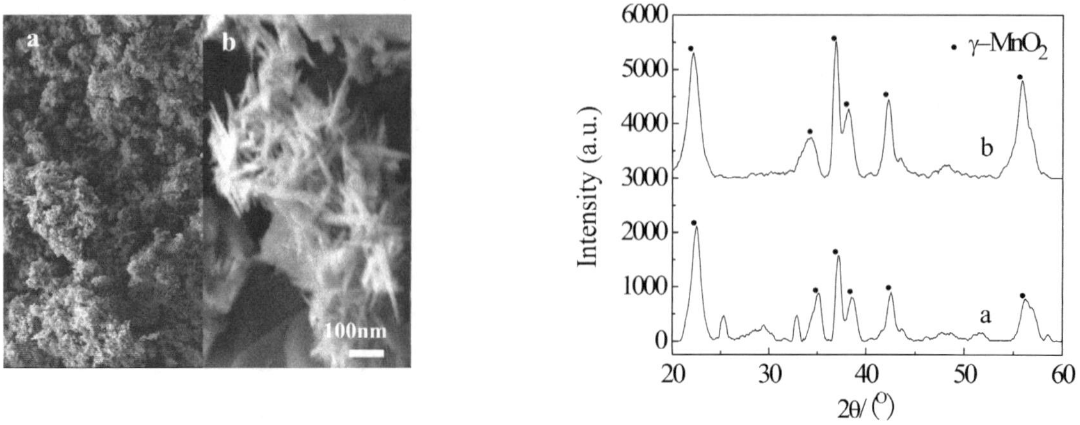

Figure 3. Influence of initial solution pH on morphology and XRD pattern of γ-MnO_2
Initial solution pH: a: 6.0-6.5, b: 3.5-4.0.

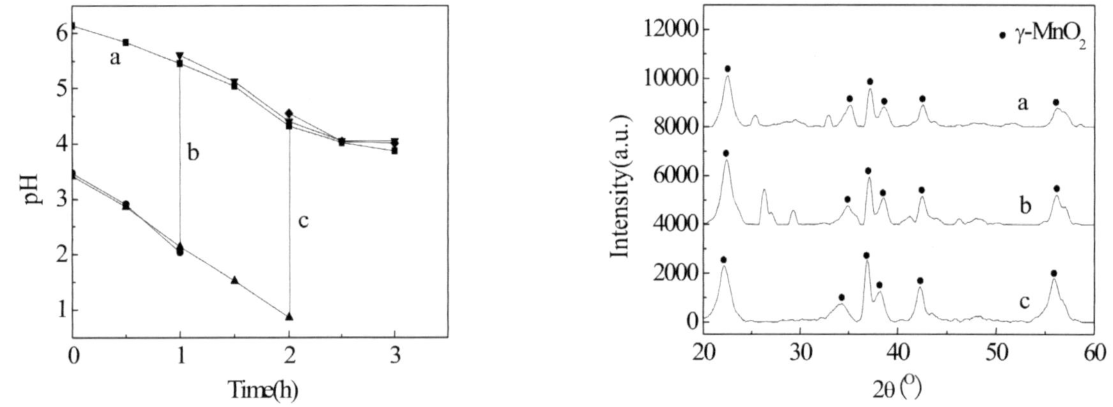

Figure 4. Variation of solution pH and XRD patterns of γ-MnO_2

3.2. Adsorption of Zn^{2+} on γ-MnO_2

Figure 5 shows the FT-IR Spectra of the γ-MnO_2 products formed at the initial solution of 6.5 and 3.5, respectively. The strong and wide peaks located around 3650-3500cm^{-1} were connected with the existence of -OH, which can be acted as the adsorption sites for the metal ions. The Zn^{2+} adsorption capacities of the γ-MnO_2 formed at the initial solution pH of 6.0-6.5 and 3.5-4.0 were 527.4 mg.g^{-1} and 131.8 mg.g^{-1}, respectively, and the difference in the adsorption capacities was considered to be connected with the varying crystallinity of γ-MnO_2.

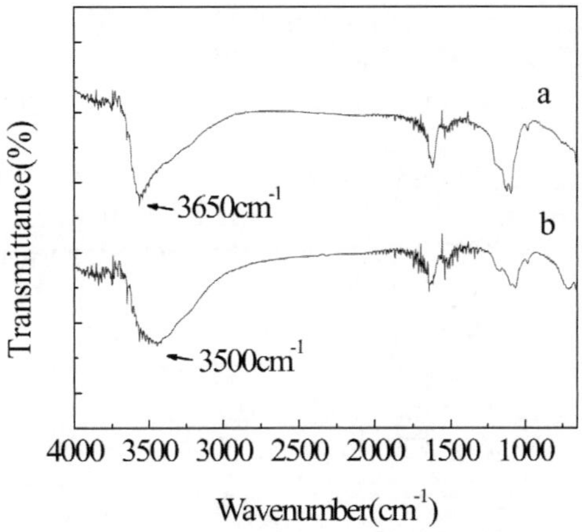

Figure 5. IR patterns of γ-MnO_2 formed at the initial solution of 6.0(a) and 3.5(b)

Table 2 shows the influence of the initial and final solution pH on the Zn^{2+} adsorption capacity of γ-MnO_2 .In the case of the final solution pH 4.0-4.1, the increase of the initial solution pH from 0.54 to 6.15 led to the increase of Zn^{2+} adsorption capacity from 131.8 mg.g^{-1} to 527.4 mg.g^{-1}, due to the poorer crystallinity of the γ-MnO_2 formed at higher initial solution pH.

Table 2. Influence of solution pH on the Zn^{2+} adsorption capacity of γ-MnO_2

Initial pH	Final pH	Adsorption capacity/ mg.g^{-1}	Initial pH	Final pH	Adsorption capacity/ mg.g^{-1}
6.15	4.10	527.4	4.50	0.55	425.1
4.50	4.03	433.6	4.50	2.07	430.2
0.54	4.05	131.8	4.51	4.45	434.3

4. Conclusions

A facile method was developed to synthesis γ-MnO_2 by the reaction of $Na_2S_2O_8$ with Zn-bearing $MnSO_4$, keeping the molar ratio of Mn to S as 1:2-4. The increase of the temperature or the final solution pH favored the oxidation of Mn^{2+} to γ-MnO_2. The crystallinity of γ-MnO_2 was connected closely with the initial solution pH. Keeping the final solution pH as 4.0-4.1, the Zn^{2+} adsorption capacities of γ-MnO_2 formed at the initial solution pH 0.54 and 6.15 were 131.8 mg.g^{-1} and 527.4 mg.g^{-1}, respectively, which was attributed to the poorer crystallinity of the γ-MnO_2 formed at higher initial solution pH.

Acknowledgments
This work was supported by the National Natural Science Foundation of China (No. 50874066).

References

[1] Suib S 2008 J. Mater.Chem. **18** 1623-31
[2] Jouanneau S, Sarciaus S, La Salle A and Guyomard D 2001 Solid State Ionics **140** 223-32
[3] Xun W and Yadong L 2002 J. Am. Chem. Soc. **124** 2880

[4] Yanming W, Haifeng L, Mi B, Binjia L, Haifeng S, Yanxuan W and Fan W 2011 Journal of Alloys and Compounds **509** 8306-12
[5] Jian Z, Shurong H, Yong Z, Lizhen W and Aiqin Z 2010 Inorganic Chemicals Industry **42** 26-8
[6] XY W, X W, WG H 2005 J. Mater. Sci. Eng. **23** 112
[7] Fangyi C, Jianzhi Z, Wene S, Chunsheng L, Hua M, Jun C and Panwen S 2006 Inorg. Chem. **45** 2038
[8] ZY Y, TZ R, GH D and BL S 2005 Appl. Phys. A: Mater. Sci. Process. **80** 743
[9] Ma R, Bando Y, Zhang T and Sasaki T 2004 Adv. Mater. **16** 918-22
[10] Deshan Z, Sixiu S, Weiliu F, Haiyun Y, Chunhua F, Guangxiang C, Zhilei Y and Xinyu S 2005 J. Phys. Chem. B **109** 16439-43
[11] Xuchun S, Yang Z and Yifan Z 2007 Cryst. Growth Des. **7** 159-62
[12] Jianzhi Z, Zhanliang T, Jing L and Jun C 2008 Cryst. Growth Des. **8** 2779-805
[13] Zhidong C, Lan G, Jianyu C, Wenchang W and Juan X 2011 ACTA CHIMICA SINICA **69** 503-7
[14] Yingpin L, Xiaoquan Z, Huijing Z, Zhurui S and Tiehong C 2007 Chemical Journal of Chinese Universities **28** 1223-26

Application of doped rare-earth oxide TiO_2:(Tm^{3+}, Yb^{3+}) in dye-sensitized solar cells

Guixiang Xie[1,2], Yuelin Wei[1], Leqing Fan[1] and Jihuai Wu*
[1] Eng. Res. Center of Environment-Friendly Functional Mater., Ministry of Education, Inst. of Materials Physical Chemistry, Huaqiao University, Quanzhou, 362021, China
[2] Longyan University, Longyan, Fujian, 364012, Chian

*E-mail: jhwu@hqu.edu.cn; lqfan@hqu.edu.cn

Received 4 August 2011, Revised 16 November 2011

Abstract. Tm^{3+} and Yb^{3+} are codoped into TiO_2 film in a dye-sensitized solar cell (DSSC). The emission and excitation spectra of TiO_2:(Tm^{3+}, Yb^{3+}) power shows that the rare-earth ions possess up-conversion luminescence function, which results in harvesting more incident light and increasing photocurrent for the DSSC. On the other hand, owing to the p-type doping effect by Tm^{3+} and Yb^{3+}, the photovoltage of the DSSC is enhanced. Under a simulated solar light irradiation of 100 mW·cm^{-2}, a DSSC containing Tm^{3+}/Yb^{3+} achieves a conversion efficiency of 7.05 %, which is increased by 10.0% compared with a DSSC lacking Tm^{3+}/Yb^{3+}.

1. Introduction

Considerable effort has been devoted to the development of dye-sensitized solar cells (DSSCs) since its first prototype was reported by Gratzel and O'Regan in 1991 [1], however, how to enhance its efficiency still is a crucial problem. To improve incident light harvest, and thus increase the photocurrent of the DSSC, many dyes have been synthesized, however, even the best of these (N-719, N-749) only absorb visible light in the wavelength range 300-800 nm [2,3], so most of the solar infrared irradiations are not utilized. If the infrared irradiations can be transferred to visible light via upconversion luminescence by rare-earth compounds, and are reabsorbed by the dye in the DSSC, more solar irradiation may be utilized, the DSSC photocurrent will be effectively enhanced.

On the other hand, the DSSC photovoltage depends on the energy level of the electron in the photoanode [2,3]. Rare-earth ions are positive trivalent cations, when they are doped in semiconductor TiO_2, a p-type doping effect occur [4,5], which results in the elevation of electron energy level of the photoanode, and turn to the enhancement of the photovoltage of the DSSC. However, little significant research on conversion luminescence and/or p-type doping by rare-earth ions in the DSSC has been attempted [6-8]. Here, Tm^{3+} and Yb^{3+}, as luminescence medium and p-type dopant, are introduced into the DSSC, it is expected that the photocurrent, photovoltage and solar conversion efficiency of the DSSC can be improved.

2. Experimental part

2.1. Materials

Yb_2O_3, Tm_2O_3 and other reagents were analytical grade (99.99%) and purchased from Shanghai Chemical Agent Ltd. China. $Ti(OC_4H_9)_4$ was purchased from Fluka. The sensitized dye N-719 was purchased from SOLARONIX SA. The above reagents were used without further purification. Conducting glass plate (FTO glass, sheet resistance 8 $\Omega \cdot cm^{-2}$) purchased from Hartford Glass Co., was cut into 2×1.5 cm^2 sheets, and was used as a substrate to precipitate a TiO_2 porous film.

2.2. Preparation of TiO_2:(Tm^{3+}, Yb^{3+}) power

TiO_2:(Tm^{3+}, Yb^{3+}) powder was prepared by the following processes: 0.056 g Tm_2O_3 and 0.58 g Yb_2O_3 were dissolved in a HNO_3 solution at 100 °C, 10 mL $Ti(OC_4H_9)_4$ was added into 150 ml distilled water under vigorously stirring. Then the above two solutions were mixed and the pH value of the mixed solution was adjusted at 2 by adding acetic acid and nitric acid solution. After peptizing at 80 °C for 12 h, the mixture was transferred to an autoclave and hydrothermally treated at 200 °C for 12 h to form a white suspension. The resultant white solid product was dried in a vacuum oven at 60 °C for 12 h. After the solid product was annealed at 800 °C for 2.5 h, a TiO_2:(Tm^{3+}, Yb^{3+}) powder was obtained.

2.3. Preparation of film electrodes and DSSCs

A TiO_2 colloid was prepared according to the methods previously used [9-11]. The TiO_2:(Tm^{3+}, Yb^{3+}) powder was dispersed in the TiO_2 colloid to form a TiO_2:(Tm^{3+}, Yb^{3+}) colloid. The TiO_2 colloid was spread on the FTO glass plate by using a doctor-blading technique. The process was done for two times to form a thick TiO_2 film (about 10 μm). The TiO_2 film was sintered at 300 ºC in air for 30 min. Then the TiO_2:(Tm^{3+}, Yb^{3+}) layer with a thickness of 5 μm was spread onto the TiO_2 film by the same method. After sintering at 450 ºC for 30 min and soaking in a 3×10^{-4} M ethanol solution of N719 for 24 h to absorb the dye adequately, a dye-sensitized rare-earth-doped TiO_2 film electrode was obtained. For comparison, a dye-sensitized TiO_2 film electrode without the rare-earth-doping (RE_2O_3) also was prepared. The thicknesses of the TiO_2 film electrodes with and without RE_2O_3 were both 15 μm.

A DSSC was assembled by injection of an electrolyte [9-11] into the aperture between the TiO_2 film electrode (anode electrode) and a platinized conducting glass electrode (cathode electrode). The two electrodes were clipped together and a cyanoacrylate adhesive was used as sealant to prevent the electrolyte solution from leaking.

2.4. Measurement and characterization

The powder XRD pattern was recorded with a X-ray diffractometer (BRUKER D8 advance, Germany) using Cu Kα radiation (λ = 1.5405 Å). The excitation and emission spectra were recorded with a spectrophotometer (FSP920, EDINBUERGH Instruments, Britain) equipped with a diode laser with wavelength 980 nm as excitation source. The photovoltaic test of DSSC was carried out by measuring the *I–V* characteristic curves under a white light irradiation from a 100 W Xe arc lamp (XQ-500W, Shanghai Photoelectricity Device Company, China) in ambient atmosphere.

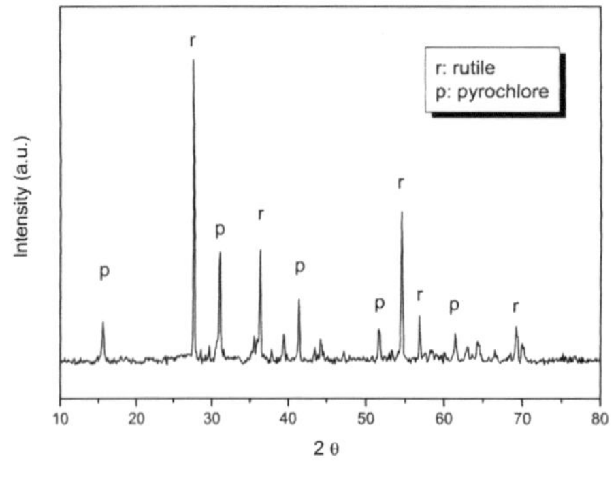

Figure 1. XRD patterns of TiO_2:(Tm^{3+}, Yb^{3+}) powder.

3. Results and discussion

3.1. XRD

Figure 1 shows the XRD patterns of TiO_2:(Tm^{3+}, Yb^{3+}) powder. The primary components in the powder are $Tm_2Ti_2O_7$ and $Yb_2Ti_2O_7$ pyrochlore phases (marked p), and TiO_2 rutile phase (marked r). According to the XRD and the Scherrer formula, the particle size of the powder is calculated to be 30 nm, this nanometer crystal size favors [12] the light collection and photoluminescence of the powder.

3.2. The excitation and emission spectra of TiO_2:(Tm^{3+}, Yb^{3+})

Figure 2a shows the excitation spectrum of TiO_2:(Tm^{3+}, Yb^{3+}). The curve shows a strong and broad excitation band in the wavelength range of 950-1000 nm. The band centered at 980 nm can be identified as the characteristic transition of Yb^{3+} ion from the ground multiplet $^2F_{7/2}$ to the excited multiplet $^2F_{5/2}$ [13,14]. This infrared excitation band with emission wavelength at 475 nm indicates that the infrared irradiation from the sun can be transferred to the visible light by TiO_2:(Tm^{3+}, Yb^{3+}).

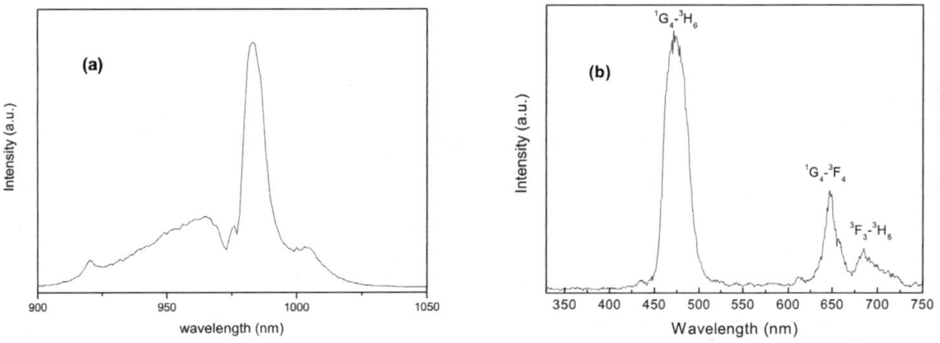

Figure 2. The excitation (a) and emission (b) spectra of TiO_2:(Tm^{3+}, Yb^{3+}) powder.

The emission spectrum of TiO_2:(Tm^{3+}, Yb^{3+}) excited with a 980-nm diode laser operating at 100 mW is shown in Figure 2b. From the Figure 2b, an up-conversion luminescence is observed, the emission band centered at 475, 647 and 685 nm are corresponded to electronic transitions for Tm^{3+} ion: $^1G_4 \rightarrow {}^3H_6$, $^1G_4 \rightarrow {}^3F_4$, $^3F_3 \rightarrow {}^3H_6$, respectively [13,14]. The luminescence emission bands for TiO_2:(Tm^{3+}, Yb^{3+}) at 475, 647 and 685 nm is just the absorption wavelength range of the dye N-719 in the DSSC. Combining the excitation with emission spectra of TiO_2:(Tm^{3+}, Yb^{3+}), the infrared irradiation from the sun can be reabsorbed by the dye N-719 in the DSSC via an up-conversion luminescence of TiO_2:(Tm^{3+}, Yb^{3+}) and the solar light harvest of DSSC can be improved.

3.3. The photovoltaic performance of DSSCs

Under a simulated solar light irradiation of 100 mW·cm^{-2}, the photovoltaic parameters of DSSCs with different doped ratio of RE_2O_3/TiO_2 in TiO_2 electrodes were measured and shown in Table 1. With the increase of the amount of RE_2O_3 in the DSSC, J_{SC} increases and then decreases. The increase in J_{SC} with the amount of RE_2O_3 at low concentrations is due to the up-conversion luminescence of RE ions, which RE ions absorb infrared light and transfers to visible light, resulting in an increasing photocurrent in the DSSC. However, when Tm^{3+} and Yb^{3+} are doped in TiO_2, some crystal defects are produced, especially in the higher concentration of RE ions. The defects can capture photoinduced electrons and holes, leading to a decrease in photocurrent [15,16]. This explains why J_{SC} increases and then decreases with the increase of the amount of RE_2O_3 in the DSSC.

On the other hand, the V_{OC} increases with increase in the percentage of RE_2O_3. According to Gratzel [2,3], V_{OC} corresponds to the difference between the Fermi level of the electrons in the oxide film and the redox potential of the electrolyte. When RE^{3+} ions are doped in TiO_2 film and substituted for the Ti^{4+} ion lattice sites, a p-type doping effect occurs, similarly to that trivalent positive ion doped in Si semiconductors, which results in the elevation of the energy level of oxide film and the increase of V_{OC}. Similar phenomena also was observed by Ko et al. [15]. They found that Al-doped TiO_2 electrode increased V_{OC}, and W-doped TiO_2 had an opposite effect (n-type doping). Although the explanation for phenomena by them is not based on p-type or n-type doping.

Comprising J_{SC}, V_{OC}, FF and η, when the RE_2O_3 amount is 3.33 wt. % in the doping layer, the light-to-electric conversion efficiency of the DSSC reaches 7.05 %, which is enhanced by 10.0 % compared with the DSSC lacking of RE ions.

Table 1. The effects of TiO_2:(Tm^{3+}, Yb^{3+}) on the photovoltaic parameters of DSSCs

RE_2O_3 (wt.% in doping layer)	J_{SC} ($mA \cdot cm^{-2}$)	V_{OC} (mV)	FF	η (%)
0.00	15.40	717	0.580	6.41
1.67	15.75	721	0.575	6.53
2.50	16.03	725	0.596	6.93
3.00	15.83	731	0.601	6.95
3.33	15.52	733	0.619	7.05
3.57	15.02	741	0.614	6.83
3.75	14.11	748	0.608	6.42

4. Conclusion

In summary, we introduce Tm^{3+} and Yb^{3+} into the TiO_2 film in dye-sensitized solar cell (DSSC). As luminescence medium, Tm^{3+}/Yb^{3+} improve the incident light harvest via an up-conversion luminescence process, transferring infrared light into visible light, enhancing the photocurrent of the DSSC. As p-type doping, the rare-earth ions elevate the Fermi level of the oxide film, heightening the photovoltage of the DSSC. Under a simulated solar light irradiation of 100 $mW \cdot cm^{-2}$, the DSSC containing rare-earth ions achieves a light-to-electric energy conversion efficiency of 7.05 %, which is enhanced by 10.0% compared with the DSSC lacking the rare-earth ions. The present findings demonstrate the feasibility of the conversion luminescence and p-type doping in DSSC and provide an effective way to improve their solar conversion efficiency.

Acknowledgement

The authors thank the support by the National High Technology Research and Development Program of China (No.2009AA03Z217), and the National Natural Science Foundation of China (No.90922028).

References

[1] O'Regan B and Gratzel M 1991 *Nature* **353** 737.
[2] Gratzel M 2009 *Acc. Chem. Res.* **42** 1788.
[3] Hagfeldt A, Boschloo G, Sun L, Kloo L and Pettersson H 2010 *Chem. Rev.* **110** 6595.
[4] Liau L and Lin C 2008 *Thin Solid Films* **516** 1998.
[5] Liau L and Lin C 2008 *Applied Surface Science* **253** 8798.
[6] Wu J, Xie G, Li S, Hao S, Lan Z, Lin J and Huang M 2010 *J. Power Sources* **195** 6937.
[7] Hafez H, Wu J, Lan Z, Li Q, Xie G, Lin J and Abdel-Mottaleb M 2010 *Nanotechnology* **21** 415201.
[8] Li Q, Lin J, Wu J, Lan Z, Wang Y, Peng F and Huang M 2011 *Electrochimica Acta* **56** 4980.
[9] Wu J, Hao S, Lan Z, Lin J, Huang M, Huang Y, Li P and Sato T 2008 *J. Am. Chem. Soc.* **130** 11568.
[10] Wu J, Hao S, Lin J, Huang M, Huang Y, Lan Z and Li P 2008 *Crystal Growth & Design* **8** 247.
[11] Lan Z, Wu J, Lin J, Huang M, Yin S and Sato T 2007 *Electrochimica Acta* **52** 6673.
[12] Wang W, Widiyastuti W, Ogi T and Lenggoro I 2007 *Chem. Mater.* **19** 1723.
[13] Boyer J, Cuccia L and Capobianco J 2007 *Nano Lett.* **7** 847.
[14] Giri N, Singh A, Rai D and Rai S 2008 *Opt. Commun.* **281** 3547.
[15] Ko K, Lee Y and Y Jung 2005 *J. Colloid and Interface Sci.* **283** 482.
[16] Murakoshi K, Kano G, Wada Y, Yanagida S and Miyazaki H 1995 *J. Electroanal. Chem.* **396** 27.

Thermal degradation kinetics of poly {N-[(4-bromo- 3,5-difluorine)-phenyl]maleimide-co-styrene} in nitrogen

Yanxun Li, Li Xu, Xinding Yao, Tingliang Luo, Guoji Liu*
College of Chemical and Energy Engineering, Zhengzhou University, Zhengzhou 450001, P.R. of china

*E-mail: guojiliu@zzu.edu.cn

Received 19 July 2011, revised 1 December 2011

Abstract. The nonisothermal degradation kinetics of the copolymer poly{N-[(4-bromo-3,5-difluorine)phenyl]maleimide-*co*-styrene}, which was synthesized with N-[(4-bromo-3,5-difluorine)phenyl]maleimide and styrene, were studied by thermogravimetry/derivative thermogravimetry techniques. The kinetic parameters, including the activation energy and the pre-exponential factor of the copolymer degradation process, were calculated by the Kissinger and Flynn-Wall-Ozawa methods. The thermal degradation mechanism of the copolymer was also studied with the Satava-Sestak method. The results show that the activation energies were 165.97 kJ/mol with the Kissinger method and 169.04 kJ/mol with the Flynn-Wall-Ozawa method. The degradation of the copolymer followed a kinetic model of a nucleation and growth and the kinetic equation could be expressed as $G(\alpha)= [-\ln(1-\alpha)]^{1/2}$ [where $G(\alpha)$ is the integral function of conversion and α is the extent of conversion of the reactant decomposed at time t]. The reaction order was 1/2.

1. Introduction

At present there is a considerable interest not only in the synthesis of new type of the plastic materials, but also in the modification of exiting polymers in order to vary its properties to meet requirements for new applications. One of the existing methods of improving polymer properties is the copolymerization. Polystyrene (PS) is a clear, colorless polymer used extensively for low-cost applications. It is available commercially in both pellet and sheet form. But there are some serious deficiencies such as low impact strength, poor weatherability and poor chemical resistance. So it is important to correct these shortcomings by modification.

N-substituted phenyl maleimide is a type of heat-resistant monomer which has a planar five-ring. The heat resistance can be improved greatly if it is embedded in the polymer chain to increase the internal rotation resistance of the chain [1-3]. As a result, it is mainly used as heat-resistant polymer modifier [4-6]. When it is adopted as the modifier, not only the heat resistance can be improved, but also processability, hot melting and abrasion resistance property will be enhanced. Furthermore, with the introducing of fluorine atom or bromine atoms, fluorine- and bromine-substituted phenyl maleimide could be obtained. Due to the fluorine atom in the polymer chain, a series of properties of the fluorine-substituted phenyl maleimide polymers, such as the thermal stability, weatherability, oxidation resistance, corrosion resistance, water resistance, oil resistance and abrasion resistance, can be improved [7,8]; Bromine-substituted phenyl maleimide is regarded as a kind of monomer with good heat resistance and flame-retardant which is used as heat-resistant polymer modifier widely [9].

In this article, we synthesized a novel heat-resistant copolymer poly{N-[(4-bromo-3,5-difluorine)phenyl]maleimide-co-styrene} with N-[(4-bromo-3,5-difluorine)phenyl]maleimide(BDPM) and styrene, and its thermal stability was investigated with thermogravimetric analysis (TGA) under dynamic conditions under a N_2 atmosphere. The thermal stability was a crucial factor for determining the processing method and the application of the copolymer.

TGA has been widely used for rapidly assessing the thermal stability of various substances, including polymer pyrolysis [10,11]. The kinetic parameters of the decomposition process, such as the rate constants, activation energies, reaction orders, and pre-exponential factors, were assessed with the data from thermograms. Many kinetic analytical methods have been established for the analysis of the TGA data [12-15]. We used the Kissinger and Flynn-Wall-Ozawa (FWO) methods to calculate the activation energy and the pre-exponential factor of the copolymer and the Satava-Sestak method to study the thermal degradation mechanism of the copolymer.

2. Materials and methods

2.1. Materials

2.1.1. Synthesis of N-[(4-bromo-3,5-difluorine)-phenyl]maleimide.

The N-[(4-bromo-3,5-difluorine)-phenyl]maleimide was prepared according to our previous report [16]: 80 ml toluene, 15 ml DMF, and 4.54 g maleic anhydride were put into a 250 ml 4-mouth flask respectively, then 8.00 g 4-bromo-3,5-difluoroaniline was added to three flasks in batches, the mixture was maintained for 1.5 h at the temperature of 293.15 K. Then, 0.88 g of toluene-p-sulfonic acid and an appropriate amount of inhibitor were added to the reaction flask, the mixture was maintained for hours at the temperature of 383.15 K until there is no water drops appear in the water separator. Toluene was removed by distillation, and then cooled to room temperature, N-[(4-bromo-3,5-difluorine)-phenyl]maleimide were obtained by filtering, washing and drying.

2.1.2. Synthesis of poly{N-[(4-bromo-3,5-difluorine)phenyl]maleimide-co-styrene}.

The copolymer poly{N-[(4-bromo-3,5-difluorine)phenyl]maleimide-co-styrene} was prepared by ourselves as following method: Styrene and BDPM were added to a 50 mL, double-jacketed glass reactor equipped with a magnetic stirrer. Azobisisobutyronitrile was dissolved in 1,4-Dioxane and then added to the flask. The reaction was kept at 70°C for 5 h to obtain a semitransparent emulsion, the whole reaction mixture was poured into methanol, and the polymer was isolated by filtration, dissolved in 1,4-Dioxane, reprecipitated, to remove the unreacted monomer and the homopolymer byproduct, and dried in vacuum at 60°C for 24 h to obtain a white powder.

2.2. FTIR measurements

Fourier transform infrared (FTIR) spectra was recorded on a FTIR Nicolet IMP410 spectrometer (America). The sample was prepared in KBr pellets, and the spectrum was obtained in the range 400-4000 cm^{-1}.

2.3. TG and DTG measurements

The TG and derivative thermogravimetry (DTG) analyses for the copolymer were conducted on a Netzsch STA 409 PG/PC thermal analyzer (Germany) with a heating rate of 10°C/min in nitrogen. In the dynamic model, heating rate of 5, 10, 15, and 20°C/min were conducted under a nitrogen atmosphere. In all the experiment, a nitrogen steam was continuously passed into the furnace at a flow rate of 30 mL/min from 40°C to 800°C. 7.5 to 8.5 mg of copolymer was analyzed in each run.

3. Results and discussion

*3.1. FTIR of poly{N-[(4-bromo-3,5-difluorine)phenyl]maleimide-*co*-styrene}*

Figure 1 shows the FTIR spectra of polystyrene and poly{N-[(4-bromo-3,5-difluorine)phenyl] maleimide-*co*-styrene}. Compared with the FTIR spectrum of polystyrene, there were strong absorptions at 1715 cm^{-1} in the FTIR spectrum of poly{N-[(4-bromo-3,5-difluorine)phenyl]maleimide-*co*-styrene}, which were the characteristic absorption peak of C=O bond. This indicated that the monomer, N-[(4-bromo-3,5-difluorine)phenyl]maleimide, was incorporated into the copolymer.

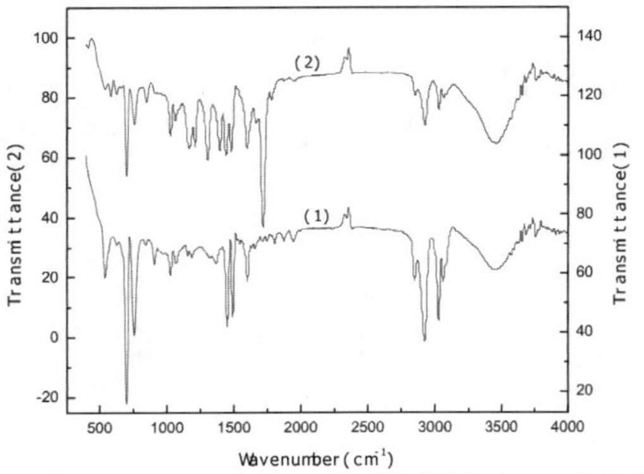

Figure 1. FTIR spectra of (1) polystyrene and (2) poly{N-[(4-bromo-3,5-difluorine)phenyl]maleimide -*co*-styrene}.

3.2. TGA and DTG curves

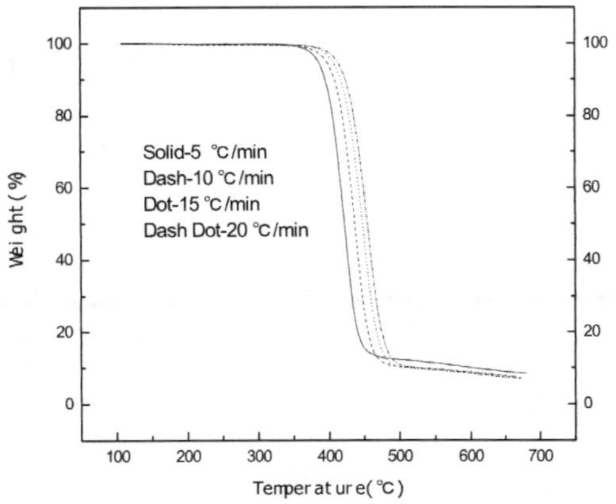

Figure 2. TGA curves of poly{N-[(4-bromo-3,5-difluorine)phenyl]maleimide-*co*-styrene} at different β values.

The TGA and DTG curves for the poly{N-[(4-bromo-3,5-difluorine)phenyl]maleimide-*co*-styrene} are illustrated in Figure 2 and 3, respectively. As shown, there was a region of major decomposition, and the TG curves shifted toward the high-temperature zone as β increased from 5 to 20°C/min because of the heat lag of the process. The decomposition behaviors at different β values were similar to one another, as indicated in Figure 3. As shown in Figure 4, there was a region of main weight loss. The main region was between 390 and 470°C with 90 wt % loss, whereas the first weight loss period of polystyrene ranged from 350 to 450°C with 100 wt % loss [17], which showed that the weight loss temperature range of poly{N-[(4-bromo-3,5-difluorine)phenyl]maleimide-*co*-styrene} was wider than that of polystyrene. This indicated that the N-[(4-bromo-3,5-difluorine)phenyl]maleimide group in poly{N-[(4-bromo-3,5-difluorine)phenyl]maleimide-*co*-styrene} made the thermal stability of polystyrene increase and delayed the thermal decomposition rate of the poly{N-[(4-bromo-3,5-difluorine)phenyl]maleimide-*co*-styrene}. This weight loss corresponded to the decomposition of the main chain. There was about 7 wt % char residue left at 650°C, which indicated a good thermal stability of the copolymer.

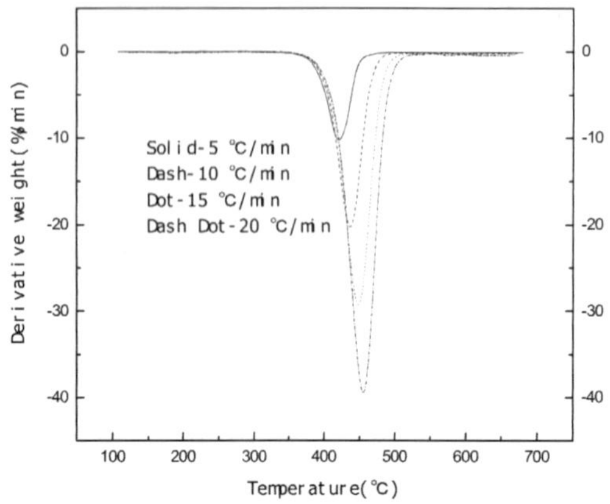

Figure 3. DTG curves of poly{N-[(4-bromo-3,5-difluorine)phenyl]maleimide-*co*-styrene} at different β values.

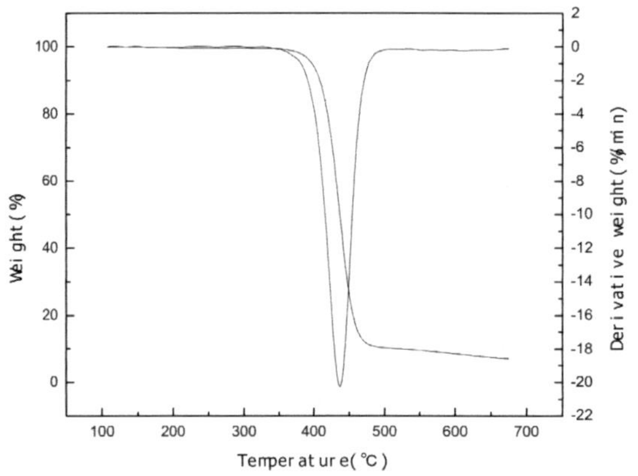

Figure 4. TGA and DTG curves of poly{N-[(4-bromo-3,5-difluorine)phenyl]maleimide-*co*-styrene} at β=10°C/min.

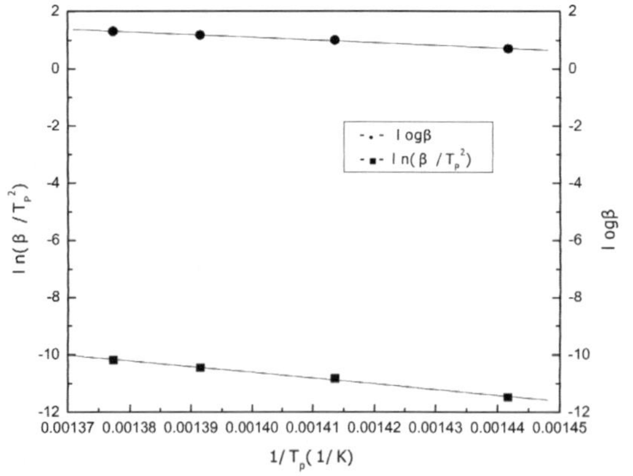

Figure 5. Plots of $\ln(\beta/T_P^2)$ versus $1/T_P$ by the Kissinger method and log β versus $1/T_P$ by the FWO method.

3.3. Theoretical background

According to the reaction theory, the kinetic equation for solid degradation, $B(s) \rightarrow D(s) + C(g)$ [where $B(s)$ is the solid reactant and $D(s)$ and $C(g)$ are the remaining solid and gas products, respectively], can usually be expressed as follows:

$$\frac{d\alpha}{dt} = kf(\alpha) \tag{1}$$

where α is the extent of conversion of $B(s)$ decomposed at time t, $f(\alpha)$ is the reaction mechanism function, and k is the reaction rate constant. k obeys an Arrhenius equation:

$$k = A\exp(-E/RT) \tag{2}$$

where A is the pre-exponential factor, E is the apparent activation energy, R is the gas constant, and T is the absolute temperature. Combining eqs. (1) and (2) gives the following equation:

$$\frac{d\alpha}{dt} = A\exp(-E/RT)f(\alpha) \tag{3}$$

If the temperature of the sample is controlled at a constant heating rate ($\beta = dT/dt$), the reaction rate can be defined as follows:

$$\frac{d\alpha}{dt} = (A/\beta)\exp(-E/RT)f(\alpha) \tag{4}$$

There are many methods for studying the kinetics of the degradation of solid materials. In this study, we used the Kissinger method [12], FWO method [13,14], and Satava-Sestak method [15] to study the degradation kinetics of the copolymer.

Through separating the variable and rearranging with the integral or differential functions of eq. (4), the Kissinger equation [eq. (5)] and the FWO equation [eq. (6)] can be gained, respectively, as follows:

$$\ln\frac{\beta}{T_p^2} = \ln\frac{AR}{E} - \frac{E}{RT_p} \tag{5}$$

$$\log(\beta) = \log(\frac{AE}{RG(\alpha)}) - 2.315 - 0.4567\frac{E}{RT} \tag{6}$$

where $G(\alpha)$ is the integral function of conversion and T_P is the temperature at the maximum weight loss. In this method, four or more thermal degradation curves are used, and the activation energy and the pre-exponential factor can be determined from the slope and the ordinate of the linear plot of $\ln(\beta/T_P^2)$ versus $1/T_P$, respectively. The Kissinger method is suitable for computing the kinetic parameters when the thermal degradation rate is fast.

Equation (6) is one of the integral methods that can be used to determine the activation energy without knowledge of the reaction order or the reaction mechanism. It is a relatively simple method for determining the activation energy directly from data of the weight loss versus the temperature obtained at several β values. The activation energy can be determined from the slope of the linear plot of ln β versus $1/T_P$. Compared to other kinetic methods, the FWO method avoids possible error resulting from different assumptions of the reaction mechanism function. Therefore, it can be used to validate the activation energy computed by different assumptions of the reaction mechanism function. It can be used to obtain the kinetic parameters of any point on the thermogravimetry (TG) curves.

Rearranging eq. (6), we obtain the Satava–Sestak equation:

$$\log G(\alpha) = \log(\frac{A_s E_s}{R\beta}) - 2.315 - 0.4567\frac{E_s}{RT} \quad (7)$$

where A_S is the pre-exponential factor obtained from Stava-Sestak equation and E_S is the apparent activation energy obtained from Stava-Sestak equation.

Thirty types of kinetic model functions [18] were used in the Satava-Sestak method. The E_S, A_S, and linear correlation coefficient (r) values of different model functions were calculated from a plot of log $G(\alpha)$ against $1/T$. It could also be used to obtain the kinetic parameters of any point on the TG curves.

3.4. Kinetic analysis

3.4.1. Evaluation of activation energy.

The Kissinger method is suitable for computing the kinetic parameters of a region at which the thermal degradation rate is fast. The FWO and Satava–Sestak methods can be used to calculate the kinetic parameters of any region in the TG curves. Table 1 shows the parameters computed from the TG curves.

Table 1. Basic Data of the Kinetics of TG Curves

β ($K \cdot min^{-1}$)	T_p(°C)	$1/T_P$ (1×10^{-3} K^{-1})	Kissinger method: $\ln(\beta/T_P^2)$ ($K^{-1} \cdot min^{-1}$)	FWO method: log β
5	420.5	1.442	-11.475	0.699
10	434.3	1.414	-10.821	1.000
15	445.5	1.391	-10.447	1.176
20	452.9	1.377	-10.180	1.301

According to Kissinger method, by plotting $\ln(\beta/T_P^2)$ versus $1/T_P$, the regression curves can be generated by the least square method, as shown in Figure 5. The apparent activation energy calculated with the Kissinger method (E_K), 165.97 kJ/mol, and the pre-exponential factor obtained with the Kissinger method, 6.75×10^{11}, were obtained. The reported activation energies of decomposition of polystyrene were 67.8 kJ/mol by other researcher [17], which were smaller than that of the copolymer. The activation energy of the polymer is determined by the energy required by the step of decomposition of the main chain. In the copolymer, the N-[(4-bromo-3,5-difluorine)phenyl]maleimide have rigid imide ring in the backbone, which mean that the decomposition of the copolymer need high temperature, this resulted in increased activation energy. In other words, to achieve decomposition, more energy was required. The result showed that the stability of the copolymer synthesized in this study was better than that of the polystyrene. With the FWO method, through the plotting of log β against $1/T_P$, the regression curves were generated with the least-square method, as shown in Figure 5; the obtained value of the apparent activation energy calculated with the FWO method (E_O), 169.04 kJ/mol, was in good agreement with E_K, and this indicated that the selection of the thermal degradation kinetics was reasonable.

3.4.2. Evaluation of the thermal degradation mechanism.

Journal of Physics: Conference Series **339** (2012) 012011 doi:10.1088/1742-6596/339/1/012011

In the Satava-Sestak equation [eq. (7)], $\log(A_S E_S/R\beta)$ is not affected by temperature. Thus, at the given value of β, a plot of log $G(\alpha)$ against $1/T$ should be a straight line with a slope of $-0.4567E_S/R$. Therefore, the function that shows a linear relation should be the fitting function. When many $G(\alpha)$ functions show the linear relation, those $G(\alpha)$ functions that meet $E \approx E_O$ should be selected. For the 30 types of thermal degradation mechanism functions (Table 2), the thermal degradation data from the TG curves at the different β values were computed by the Satava–Sestak method. As a result, the number 11 function was the most suitable kinetic model function; that is, $G(\alpha) = [-\ln(1-\alpha)]^{1/2}$. $|r|$ was larger than 0.96, and the standard deviation (SD) was less than 0.07; this indicated good linearity, as shown in Tables 1 and 3. Therefore, the obtained kinetic equation of the thermal degradation of the poly{N-[(4-bromo-3,5-difluorine)phenyl]maleimide-*co*-styrene} was $G(\alpha) = [-\ln(1-\alpha)]^{1/2}$. This showed that the degradation of the copolymer followed a kinetic model of the nucleation and growth, and the order of the reaction was 1/2.

Table 2. Thirty Types of Thermal Degradation Mechanism Functions

No.	Differential function: $f(\alpha)$	Integral function: $G(\alpha)$
1	$1/2\ \alpha^{-1}$	α^2
2	$-[\ln(1-\alpha)]^{-1}$	$\alpha + (1-\alpha)\ln(1-\alpha)$
3	$3/2[(1-\alpha)^{-1/3} - 1]^{-1}$	$(1 - 2\alpha/3) - (1-\alpha)^{2/3}$
4 and5	$3/n(1-\alpha)^{2/3}[1-(1-\alpha)^{1/3}]^{-(n-1)}$ $(n = 2, 1/2)$	$[1-(1-\alpha)^{1/3}]^n$ $(n = 2, 1/2)$
6	$4(1-\alpha)^{1/2}[1-(1-\alpha)^{1/2}]^{1/2}$	$[1-(1-\alpha)^{1/2}]^{1/2}$
7	$3/2(1+\alpha)^{2/3}[(1+\alpha)^{1/3} - 1]^{-1}$	$[(1+\alpha)^{1/3} - 1]^2$
8	$3/2(1-\alpha)^{4/3}[(1-\alpha)^{-1/3} - 1]^{-1}$	$[(1/(1+\alpha))^{1/3} - 1]^2$
9	$1-\alpha$	$-\ln(1-\alpha)$
10–16	$1/n(1-\alpha)[-\ln(1-\alpha)]^{-(n-1)}$ $(n = 2/3, 1/2, 1/3, 4, 1/4, 2, 3)$	$[-\ln(1-\alpha)]^n$ $(n = 2/3, 1/2, 1/3, 4, 1/4, 2, 3)$
17–22	$1/n(1-\alpha)^{-(n-1)}$ $(n = 1/2, 3, 2, 4, 1/3, 1/4)$	$1-(1-\alpha)^n$ $(n = 1/2, 3, 2, 4, 1/3, 1/4)$
23–27	$1/n\ \alpha^{-(n-1)}$ $(n = 1, 3/2, 1/2, 1/3, 1/4)$	$\alpha^n (n = 1, 3/2, 1/2, 1/3, 1/4)$
28	$(1-\alpha)^2$	$(1-\alpha)^{-1}$
29	$(1-\alpha)^2$	$(1-\alpha)^{-1} - 1$
30	$2(1-\alpha^{3/2}$	$(1-\alpha)^{-1/2}$

Table 3. Results of the Satava-Sestak Method

β (K/min)	E(kJ/mol)	Log A_S	r	SD
5	163.57	11.56	0.9716	0.0614
10	155.29	10.99	0.9646	0.0674
15	158.52	11.25	0.9725	0.0608
20	161.33	11.46	0.9698	0.0635

4. Conclusions

The thermal degradation process of poly{N-[(4-bromo-3,5-difluorine)phenyl]maleimide-*co*-styrene} was studied by the TG-DTG method. The results show that the activation energies obtained by the Kissinger and FWO methods were 165.97 and 169.04 kJ/mol, respectively. The degradation of the copolymer followed a kinetic model of the nucleation and growth, the kinetic equation could be expressed as $G(\alpha) = [-\ln(1-\alpha)]^{1/2}$, and the reaction order was 1/2.

Reference

[1] Mokhtar S M, Abd-Elaziz S M, Gomaa F A 2010 *Journal of Fluorine Chemistry* **131** 616
[2] Matsumoto A, Kubota T, Otsu T 1990 *Macromolecules* **23** 4508
[3] Otsu T, Matsumoto A, Kubota T, Mori S 1990 *Polym Bull* **23** 43
[4] Chung C M, Ahn K D 1997 *Macromol Symp* **118** 485
[5] Kita Y, Kishino K, Nakagawa K 1997 *J. Appl. Polym. Sci.* **63** 1055
[6] Matsumoto A, Kimura T 1998 *J. Appl. Polym. Sci.* **68** 1703
[7] Lee J R, Jin F L, Park S J, Park J M 2004 *Surf. Coat. Technol.* **180** 650
[8] Pomes V, Fernandez A, Costarramone N, Grano B, Houi D 1999 *Colloid Surf.* **159** 481
[9] Ryttel A 1999 *Angew. Makromol. Chem.* **267** 67
[10] Li L Q, Guan C X, Zhang A Q, Chen D H, Qing Z B 2004 *Polym. Degrad. Stab.* **84** 369
[11] Liu B Y, Li Y, Zhang L, Yan W D, Yao S H 2007 *J. Appl. Polym. Sci.* **103** 3003
[12] Kissinger H E 1957 *Anal. Chem.* **29** 1702
[13] Flynn J H, Wall L A 1966 *J. Polym. Sci. Part B: Polym. Lett.* **4** 323
[14] Ozawa T 1965 *Bull. Chem. Soc. Jpn.* **38** 1881
[15] Satava V, Sestak J 1975 *J. Therm. Anal.* **8** 477
[16] Li Y X, Xu L, Yao X D, Luo T L, Liu G J 2011 *J. Chem. Eng. Data* **56** 158
[17] Nishizaki H, Yoshida K, Wang J H 1980 *J. Appl. Polym. Sci.* **25** 2869
[18] Hu R Z, Shi Q Z 2001 *Thermal Analysis Kinetics* （Beijing: Science） pp 56–67

Structural analysis of TiO_2 and TiO_2-Ag thin films and their antibacterial behaviors

J H Hsieh[1,*], R B Yu[1], Y K Chang[2], C Li[3]

[1]Department of Materials Engineering and Center for Thin Film Technologies & Applications, Ming Chi University of Technology, Taishan, Taipei 24301, Taiwan
[2]Department of Chemical Engineering, Ming Chi University, Taishan, Taipei 24301, Taiwan
[3]Dept. of Mechanical Engineering, National Central University, Chun-Li, Taiwan

***E-mail**: jhhsieh@mail.mcut.edu.tw

Received 31 July 2011, Revised 29 November 2011

Abstract. TiO_2 (rutile and anatase) thin films was first prepared using reactive sputtering, in an Ar+O_2 plasma. In the 2nd stage of the experiment, various amounts (3, 7, and 10 at. %) of Ag was doped into the rutile film in order to form TiO_2-Ag thin films. These films were annealed for one hour in Ar atmosphere, at 300, 400, and 500 °C. The films' structures were then examined using X-ray diffractometry. FESEM (field-emission scaning electron microscopy) was used to investigate the surface emergence of Ag particles. As for the examination of optical band gaps and absorption of these films, UV-Vis-NIR photometer was used. The results show that, in as-deposited condition, the addition of Ag might disrupt the growth of crystalline structure and cause the formation of amorphous films. After annealing, it is found that the structure tends to become anatase phase which is a metastable phase between amorphous titanium oxide and rutile. More importantly, the absorption of the Ag-doped films would be enhanced in the visible-light range. Some of the enhancement is clearly due to plasmon resonance effect. The Ag-doped samples have shown some antibacterial effect in dark. When irradiated with light, the samples show a synergistic behavior combining the bactericidal effect of Ag ions and photocatalytic effect of TiO_2.

1. Introduction

TiO_2 attracts many researchers due to its interesting properties that include chemical stability, good optical transparency, high refractive-index, etc. Moreover, TiO_2 by far is the most popular photocatalytic material. Therefore, it can be used as an optical coating [1] as well as a good photocatalyst [2]. Unfortunately, TiO_2, although is an inexpensive and stable photocatalyst, the wide band gap nature makes the material only works under UV light irradiation. Hence, the development of highly efficient TiO_2-based photocatalysts is still a much interesting topic.

TiO_2 thin films can be prepared using various techniques that include sol-gel method [3], chemical vapor deposition [4], and physical vapor deposition [5]. Although three polymorphous phases can exist, one can normally find rutile (band gap=3.0 eV) and/or anatase (band gap=3.2 eV) phases in a thin film state. Due to the difference in band gap, anatase TiO_2 requires an irradiation light with wave length shorter than 380 nm, in order to function, while rutile phase require a wave length shorter than 413 nm. However, while researchers are still debating which phase (or a mixed phase) can be a more efficient photocatalyst, many researchers agree that it is necessary to extends TiO_2's absorption edge into visible light range. In order to make it work under visible light irradiation, many studies doped TiO_2 with N, C, F, and P elements or some transition metals. A thorough review in this regard can be found in Ref. [6]. Depending on the concentrations of various dopants, many studies show that the efficiency of the doped TiO_2 can be improved under visible light irradiation. Recently, more and more studies are focusing on embedding Ag nanoparticles into semiconductive oxide films. Ag is a nontoxic precious metal with remarkable catalytic activity [7, 8].

Published under licence by IOP Publishing Ltd

In addition, Ag particles may have promising size- and shape-dependent optical properties under visible light irradiation [9, 10].

In sum, adding Ag into an oxide-based photocatalyst may have many advantages, as described in the following:

1. The band gap may be narrowed [11].
2. Ag can be an efficient antibacterial material with or without light irradiation [12]
3. Ag doping may increase the absorption of visible light due to plasmonic effect [13]
4. Ag can serve as electron trapping sites (avoid recombination) due to Schottky barrier effect [14, 15]. It can also facilitate electron excitation by creating a local electric field.

The present study is aiming at preparing TiO_2 and Ag-doped TiO_2 thin films, and studying their structural changes as well as the variations of optical properties, before and after annealing, as functions of Ag doping and annealing temperature. The processing conditions for producing anatase and rutile films were first determined. Following this, TiO_2-Ag films were prepared and characterized. Finally, the antibacterial behavior is examined.

2. Experimental methods

TiO_2 thin films were first deposited on fused quartz glass substrates by reactive DC magnetron sputtering of a metallic Ti target. The target had a diameter of 50 mm and was tilted by 30° to the substrate holder. The distance of target-to-substrate was 100 mm. For deposition, the sputtering system was first pumped down to $7x10^{-4}$ Pa and then Ar gas (35 sccm) was introduced to fill the chamber up to 0.65 Pa. The substrates were ultrasonically cleaned in acetone (5 min) and isopropanol (5 min), then, blown dry. Prior to film deposition, the target was sputter-cleaned in Ar plasma at a pressure of 0.65 Pa for 10 min, with shutter closed. At the same time, the substrates were cleaned in Ar plasma for 10 min. by applying a bias of 100 W (RF).

The deposition time was around 30 min, while the film thickness was controlled nearby 500 nm. Film thickness was determined by a surface profiler (Kosaka Surfcoder). After the processing conditions for producing anatase and rutile TiO_2 films were determined, TiO_2–Ag thin films were prepared using reactive co-sputtering with Ti and Ag targets. In this case, the contents of Ag was controlled around 3, 7, and 10 at.% in the TiO_2-Ag films. Following the deposition, Some of the deposited films were then annealed for one hour at 300, 400, and 500 ^oC using a rapid thermal annealing (RTA) system (SJ, ARTS-150) in an argon atmosphere, with ramping rate set at 100 ^oC/sec.

The phases of the deposited films, before and after annealing, were examined by X-ray diffractometry (XRD). For the optical properties, the film's absoption (Absorption= 100%-Transmission-Reflection) was measured by a UV-VIS-NIR double beam spectrophotometer (JASCO-V750). For the investigation of the surface emergence of Ag particles, the films were examined using field-emission scanning electron microscopy (JEOL 6700F).

For the antibacterial testing, the coated and uncoated samples were cut and placed on a tray and then incubated with a calibrated bacterial suspension of E. coli. The samples were then incubated at 37 ^oC for various times up to 24 hours. After incubation, the antibacterial property could be evaluated according to the value of Antibacterial Efficiency (E) which was calculated according the following equation:

$$E\ (\%) = [(A-B)/A] \times 100\% \qquad (1)$$

where A=number of viable bacteria with uncoated (standard) sample in the tray; B=number of viable bacteria with coated sample in the tray.

3. Results and discussions

3.1 Structural analysis

Figure 1 shows the X-ray diffraction patterns of as-deposited samples. It is found that the rutile pattern is slowly destructed and replaced by Ag pattern when the amount of Ag is increased from 3 at.% to 10 at.%. Apparently, the doping of Ag during deposition would disrupt the growth of crystalline rutile structure. This is known as the second element effect [16]. Figures 2 show the comparison of X-ray diffraction patterns of TiO_2-10%Ag with and without annealing, as functions of annealing temperature and Ag contents. It is observed that the addition of Ag can enhance the formation of anatase phase after annealing. This can be explained as the following: The incorporation of Ag first disrupt the structure of rutile and cause the formation of amorphous state, as observed in Figure 1. During annealing, the metastable anatase phse can nucleate easier than rutile phase. This has become more obvious when the contents of Ag reach 7 at.% and higher. When the starting phase is an anatase phase, the observed result is similar to that of the rutile phase. The annealed TiO_2 phase tends to become anatase if Ag is doped into the films.

Figure 3 shows the surface SEM micrographs of the sample with 7at.% Ag, and was annealed at 400 °C. The emergence of Ag particles on films' surface is observed on all the Ag-doped samples.

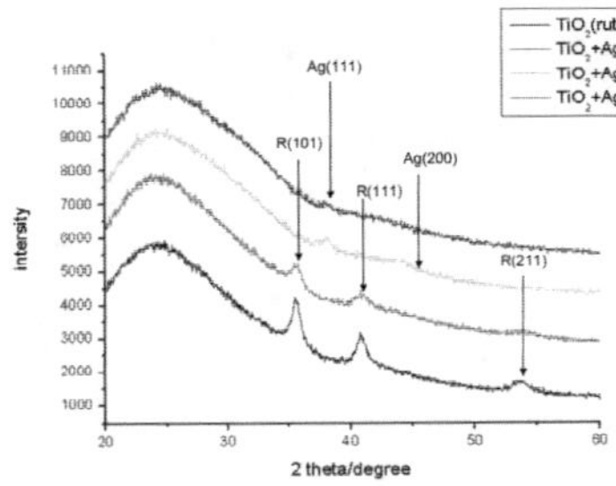

Figure 1. X-ray diffraction patterns of TiO_2(rutile) and TiO_2-Ag thin films in as-deposited states.

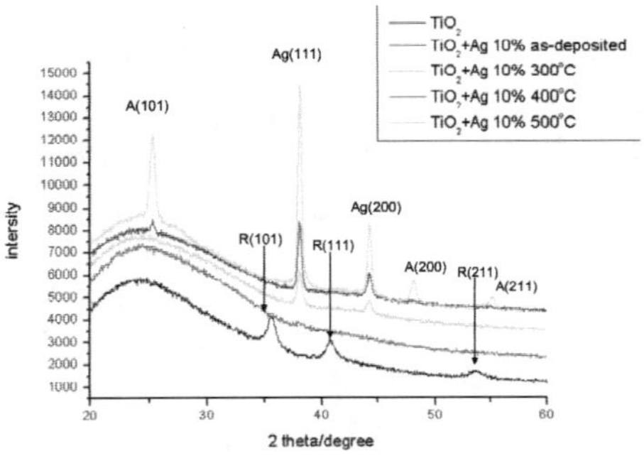

Figure 2. X-ray diffraction patterns of TiO_2-Ag (10 at.%) with and without annealing (Annealing temperature: 300°C, 400°C and 500°C)

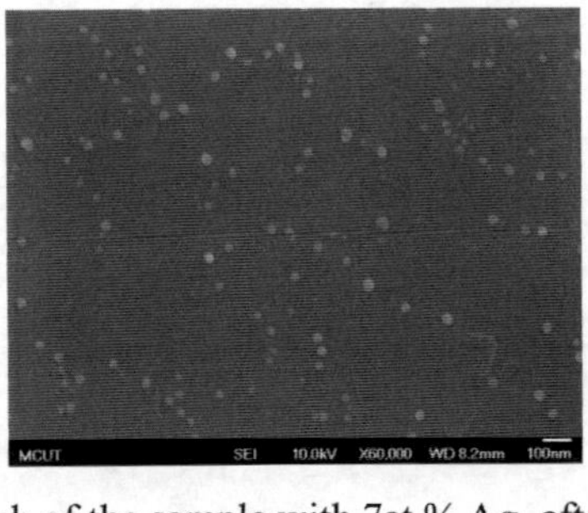

Figure 3. Surface SEM micrograph of the sample with 7at.% Ag, after being annealed at 400 °C.

3.2 Optical properties

Figure 4 shows the absorption spectra for all the as-deposited films. It is found that the sample with 7 at.% Ag has a strong and wide absorption peak between 400-550 nm. This is known due to plasmon resonance absorption. Awazu et al. [13] recently proposed a new term based on plasmon-enhanced photocatalytic effect. The new term was named "plasmonic photocatalysis". The idea of plasmonic photocatalysis utilizes the assistance of plasmon resonance that is known to enhance light absorption. For anatase TiO_2 phase to be activated, a UV irradiation with wavelength shorter than 380 nm is required to generate electron-hole pairs. In the case of Ag-doped TiO_2, the embedded Ag nano-particles could show a very intense absorption band within 400-500 nm due to the influence of plasmon resonance [17]. This is associated with a considerable enhancement of the electric near-field in the vicinity of the Ag nanoparticles. It is thought that this near-field enhancement could boost the excitation of electron-hole pairs in TiO_2 and therefore increase the efficiency of the photocatalysts. Similar related ideas were already outlined in the past [18, 19]. It is known that the resonance wavelength depends on the size and shape of the Ag nano-particles, as well as the particle-forming material. Therefore, the extended absorption peak is not observed on some other samples. Figure 5 shows the absorption spectra for the annealed TiO_2-10%Ag with the variation of annealing temperature. For most of the samples, the plasmon effect is still obvious. For the samples with 10 at.% Ag, high annealing temperature (i.e. 500 °C) can cause a full range absorption. This could be due to the formation of large Ag metal particles. From the calculation using Tauc formula, $(\alpha E)^{1/2}=(E-E_g)$, and Lambert-Beer law, it is found that the addition of Ag into TiO_2 could decrease the value of band gap form 3.2 to 2.9 eV, at maximum.

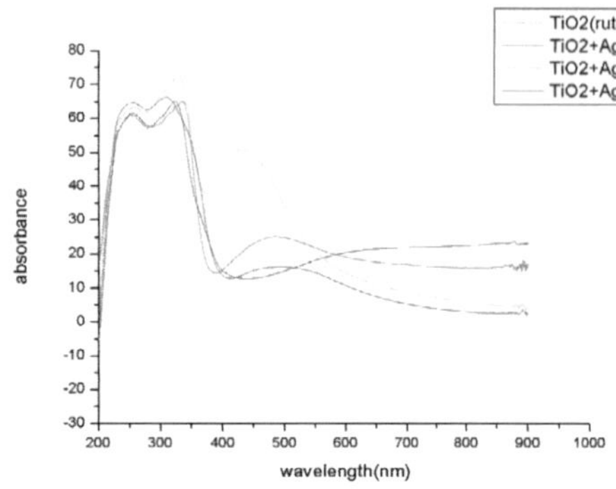

Figure 4. The absorption spectra of TiO_2 (rutile) and TiO_2-Ag thin films in as-deposited state.

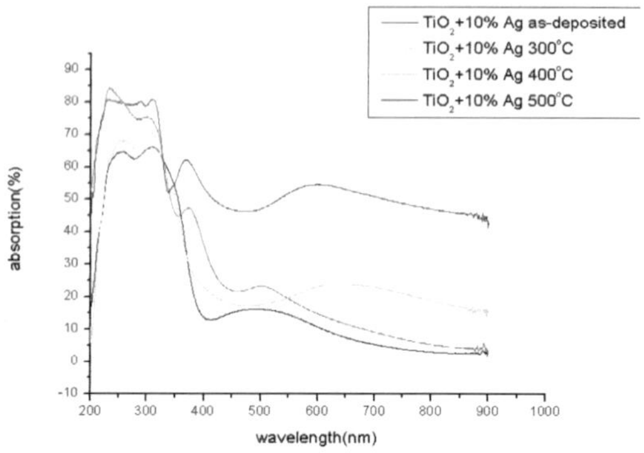

Figure 5. The absorption spectra of TiO_2-Ag (10 at.%) with and without annealing (Annealing temperature: 300°C, 400°C and 500°C)

Figure 6 shows the antibacterial efficiency of TiO_2 and TiO_2-Ag thin films tested in dark. It is obvious that the surfaced Ag can dissolve and cause the damage of E.coli. Figure 7 shows the same group of samples tested under the irradiation of visible light with peak intensity at 500 nm. According to the results, it can be observed that a synergistic effect combining Ag and photocatalytic characteristics of TiO_2 is observed. Clearly, the doped Ag can help extending the effective light absorption of TiO_2.

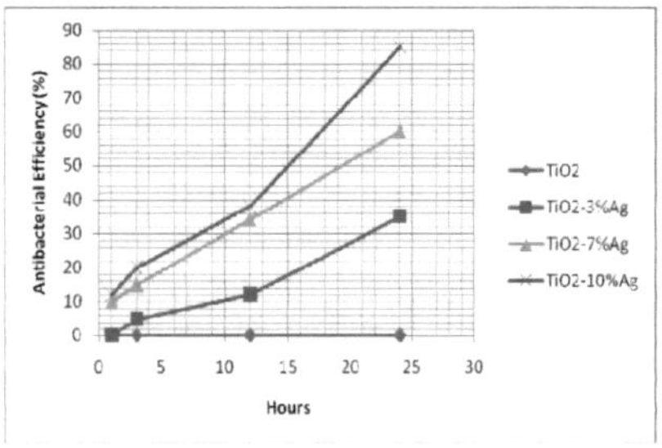

Figure 6. Antibacterial effecncy of TiO_2 and TiO_2-Ag thin films after being annealed at 500 °C for one hour. (Tested in dark)

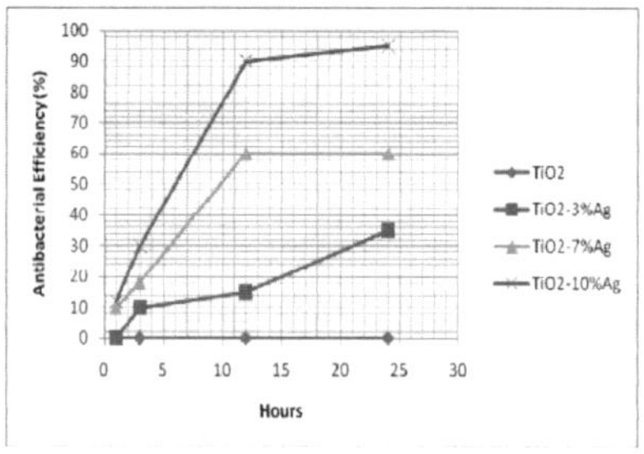

Figure 7. Antibacterial effecncy of TiO_2 and TiO_2-Ag thin films after being annealed at 500 °C for one hour. (Tested in visible light)

4. Conclusions

TiO_2 (rutile and anatase) and TiO_2-Ag thin films were prepared by reactive co-sputtering processes. The structure and optical properties were examined and analyzed as functions of annealing temperature and Ag contents. The results show that, in an as-deposited condition, the addition of Ag might disrupt the growth of crystalline structure and cause the formation of amorphous films. After annealing, it is found that the structure tends to become anatase phase, particularly when the Ag content is increased. This is due to that anatase phase is considered to be a metastable phase between amorphous state and rutile structure. More importantly, the absorption of the Ag-doped films would be enhanced in the visible-light range. Some of the enhancement is clearly due to plasmon resonance effect. This makes plasmonic photocatalysis possible. The Ag-doped samples have shown some antibacterial effect in dark. When irradiated with light, the samples show a synergistic behavior combining the effect of Ag ions and photocatalytic effect of TiO_2.

References

[1] Sawada Y and Taga Y 1984 Thin Solid Films **116** 155
[2] Norby T 1999 Solid State Ionics **125** 1
[3] Yu J and Zhao J 2002 Applied Catalysis B: Environmental **36** 31
[4] Lu J, Wang J and Raj R 1991 Thin Solid Films, **204** L13
[5] Miyaki S, Kobayashi T, Satou M, Fijimoto F 1991 Journal of Vacuum Science and Technology **A9** 3036
[6] Ekambaram S 2008 Journal of Alloys and Compounds **448** 238
[7] Guin D, Manorama S V, Latha J N L and Singh S. 2007 J. Phys. Chem. C **111** 13393
[8] Arabatzis I M, Stergiopoulos T, Bernard M C, Labou D, Neophytides S. G. and Falaras P (2003) Appl. Catal., **B 42** 187
[9] Ohko Y, Tatsuma T, Fujii T, Naoi K, Niwa C, Kubota Y and Fujishima A 2003 Nat. Mater. **2** 29
[10] Naoi K, Ohko Y and Tatsuma T 2004 J. Am. Chem. Soc. **126** 3664
[11] Page K, Palgrave R G, Parkin I P, Wilson M, Savin S L P, Chadwick A V 2007 J. Mater. Chem. **17** 95
[12] Skorb E V, Antonouskaya L I, Belyasova N A, Shchukin D G, Mohwald H and Sviridov D V 2008 Applied Catalysis B: Environmental **84** 94
[13] Awazu K, Fujimaki M, Rockstuhl C, Tominaga J, Murakami H, Ohki Y, Yoshida N and Watanabe T 2008 J. Am. Chem. Soc. **130** 1676
[14] Zhao G, Kozuka H, Yoko T 1996 Thin Solid Films **277** 147
[15] Herrmann J M, Tahiri H, Ait-Ichou Y, Lassaletta G, Gonzalez-Elipe A R, Fernandez A 1997 Appl. Catal. **B 13** 219
[16] Zeman P, Cerstvy R, Mayrhofer P H, Mitter C, Musil J 2000 Mater. Sci. Eng. **A 289** 189
[17] Kerker M 1985 J. Colloid Interface Sci. **105** 297
[18] Hirakawa T. and Kamat P V 2004 Langmuir **20** 5645
[19] Hirakawa T and Kamat P V 2005 J. Am. Chem. Soc. **127** 3928

Effect of transition metal elements addition on the properties of nitrogen-doped TiO_2 photocatalysts

H Li[1,2], P Zhang[1], S Yin[1], Y Wang[2], Q Dong[1], C Guo[1], and T Sato[1]*

[1]Institute of Multidisciplinary Research for Advanced Materials, Tohoku University, 2-1-1 Katahira, Aoba-ku, Sendai, 980-8577, Japan
[2]Lanzhou University, P.R.China

*E-mail: tsusato@tagen.tohoku.ac.jp

Received 22 October, 2011, Revised 26 November, 2011

Abstract. Nitrogen-doped TiO_2 ($TiO_{2-x}N_y$) nanoparticles with and without adding the transition metal ions, such as Co^{2+} and Ni^{2+}, were synthesized by thermohydrolysis of $TiCl_3$. The samples were characterized by X-ray diffraction, specific surface area determination, UV-vis diffuse reflectance spectroscopy. The photocatalytic activity of the sample was investigated by employing the oxidative destruction of nitric oxide as a probe reaction using a continuously flowing gases-solid system. Although the doping of transition metal ion in the lattice of titania was not useful to improve the photocatlytic activity, loading of transition metal oxides on the surface of titania resulted in an improvement of the photoreactivity of the nitrogen-doped TiO_2. The beneficial effect was explained by an increased separation efficiency of the photogenerated electron-hole pairs.

1. Introduction

Titanium dioxide is the most-used photocatalyst because of the facile synthesis, stability under various reactions media, cheapness, and bands position compatible with the oxidation of many organic compounds [1,2]. However, because of its large band gap (3.2 eV), TiO_2 can be only excited by UV light irradiation of less than the wavelength of 400 nm. Several studies have been carried out with the aim to extend the optical response to the visible-light region and to slow down the recombination rate of the electron-hole pairs by selective doping of TiO_2 with metal ions [3-5] or with non-metal anionic species [6-8]. Many controversial results are reported in literature because the method of doping strongly determines the properties of the resulting catalysts.

The most important advance in this field should be attributed to the nitrogen doped TiO_2, which was first reported by Asahi et al. in 2001 [9,10]. By doping with nitrogen ion from N_2 or NH_3 atmosphere or other nitrogenous organics, the band gap of titania could be narrowed since the valence band of N2p band locates above O2p band. While, besides anion doping, studies on transition metal ion doping have also proven to be an efficient method to improve the photocatalytic properties of TiO_2. Among them, Fe, Co, and Ni were suggested as one of the most promising candidates due to its effect in depressing electron-hole recombination rate and shifting the absorption onset to visible-light region [11-16]. In a previous paper [12], it was reported that the Fe and N co-doping titania, $Ti_{1-x}FeO_{2-y}N_z$,

Journal of Physics: Conference Series **339** (2012) 012013 doi:10.1088/1742-6596/339/1/012013

caused the decrease in photocatalytic activity, due to the formation of lattice vacancy, but $TiO_{2-x}N_y/Fe_2O_3$ composites showed excellent photocatalytic activity, due to the heterogeneous electron transfer from $TiO_{2-x}Ny$ to Fe_2O_3 to returd the quick recombination of photoinduced electrons and holes.

In this work, nitrogen-doped titania added transition metal ion, such as Co^{2+} and Ni^{2+} were prepared by two different hydrothermal methods to clarify the effect of transition metal ion addition on the photocatalytic activity.

2. Experimental

2.1. Preparation of photocatalysts

Co^{2+} or Ni^{2+} contained $TiO_{2-x}N_y$ was prepared by the similar hydrothermal method reported in a previous paper [12,17,24]. 20 wt.% of $TiCl_3$ aqueous solution and hexamethylenetetramine (HMT) were used as the titanium and nitrogen source, respectively. $CoCl_2{\cdot}6H_2O$ and $Ni(NO_3)_2{\cdot}6H_2O$ were used as Co^{2+} or Ni^{2+} sources. These ions were added to $TiO_{2-x}N_y$ by two different methods, such as the 1-step reaction and 2-step reaction. 1-step reaction: after dissolving all starting materials, the solution was heated at 190°C for 15 min by a microwave oven. 2-step reaction: After fabricating $TiO_{2-x}N_y$ nanoparticles by the microwave assisted hydrothermal reaction of $TiCl_3$ and HMT, the transition metal salt was added in the solution, followed by a hydrothermal reaction at 190°C for 15 min by a microwave oven. Finally, the samples were vacuum dried at 60°C overnight. The amount of transition metal ions in the final products was changed in the range from 1 to 10 mol.%.

2.2. Characterization

The chemical compositions of the samples were determined by the EDX spectrometer (Rayny EDX-800HS, Shimadzu). The crystalline phases of the samples were determined with X-ray diffraction analysis (XRD, Shimadzu XD-D1). The morphology of the samples was observed by TEM (JEOL, JEM-2010). BET measurements (Quantachrome Instruments, NOVA 4200e) were performed to determine the specific surface areas. The ultraviolet-visible diffuse reflectance spectrum was taken by a UV-vis spectrophotometer (Shimadzu, UV-2450).

2.3. Photocatalytic reaction

To evaluate the photocatalytic activities of the prepared samples, the oxidative decomposition behaviour of NO gas was investigated using a flow type reactor under irradiation of a 450 W high pressure mercury arc at room temperature. The detailed experimental apparatus used for the photocatalytic reaction was reported in a previous paper [18]. During the experiment, a 2 ppm standard NO gas (balance N_2) was mixed with air (1:1), and passed through the reactor (373 cm^3 of internal volume) continuously throughout the reaction. The sample was placed in the hollow (20 mm×16mm×0.5 mm) of a glass holder plate and set in the bottom center of the reactor. To reach the adsorption and desorption equilibrium, the sample was allowed to contact with the NO gas for more than 10 min. After that, the light was turned on to start the photocatalytic reaction. The mercury arc was put 165 mm away from the reactor, and it was cooled by using a Pyrex jacket with cooling water at 30 °C during the irradiation. Furthermore, an electric fan was also used to cool the reaction system. The wavelength of the irradiation light was controlled using several filters: Pyrex glass for > 290 nm, Kenko L41 Super Pro (W) filter > 400 nm and Fuji triacetyl cellulose filter >510 nm. The evaluation time for one sample under the irradiation with each wavelength was 10 min. The concentration of NO was determined using a NO_x analyzer (Yanaco, ECL-88A). It has been reported that during the photocatalytic decomposition, about 20 % of NO is directly reduced to N_2 and the other 80 % is oxidized to NO_3^- species [19].

3. Results and discussion

Table 1. EDX analysis results of the transition metal – nitrogen co-doped TiO_2 synthesis by 1-step and 2-step methods.

Added amount /mol.%	Co / mol.%		Ni / mol.%	
	1-Step	2-Step	1-Step	2-Step
1	1.2	1.1	0.9	1.3
3	3.2	2.8	3.1	3.3
5	4.2	4.1	4.3	4.1
19	9.4	9,1	8.5	8.7

The Co^{2+} and Ni^{2+} contents in the prepared samples are shown in Table 1. The results suggest that Co^{2+} or Ni^{2+} added were almost completely incorporated in the sample by both 1-step and 2-step reactions.

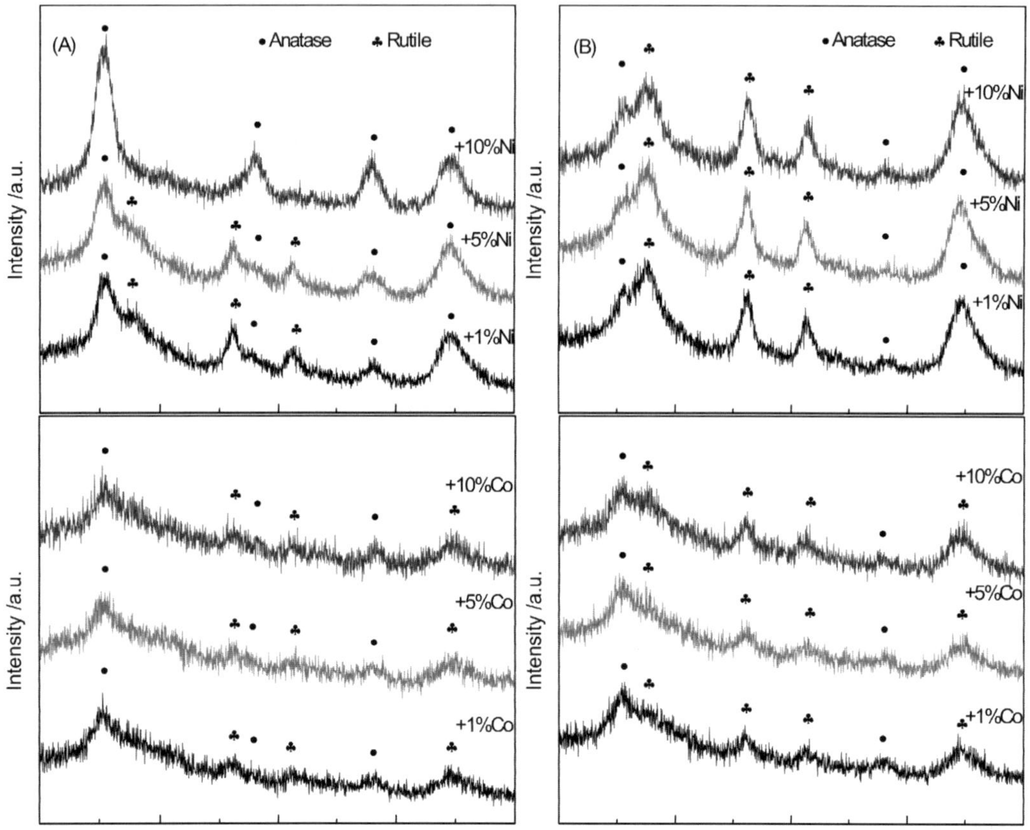

Figure 1. XRD patterns of the samples prepared by (A) 1-step method and (B) 2-step method with different Co^{2+} and Ni^{2+} contents.

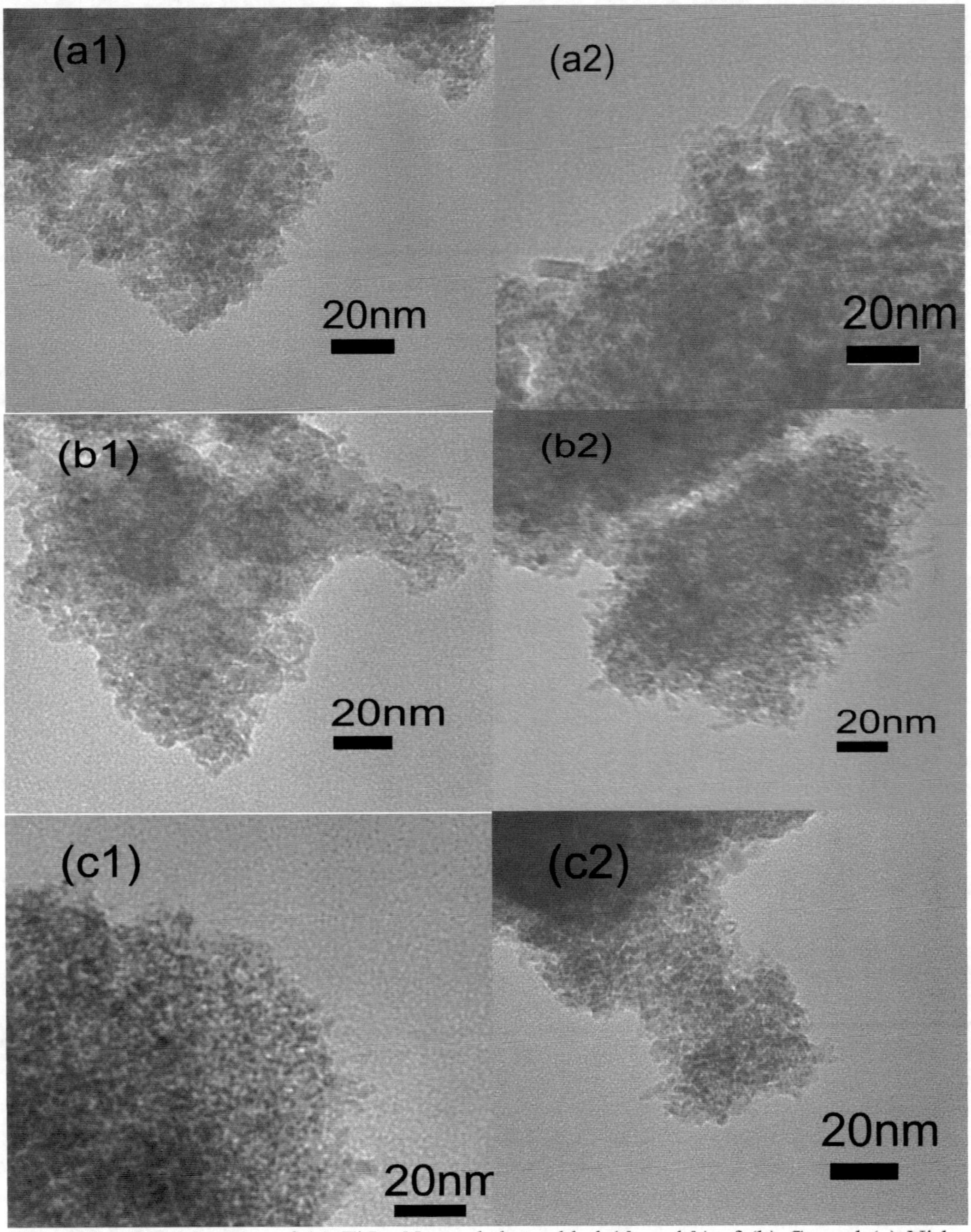

Figure 2. TEM images of (a) $TiO_{2-x}N_y$, and that added 10 mol.% of (b) Co and (c) Ni by (a1,b1,c1) 1-step and (a2,b2,c2) 2-step method.

The XRD patterns of the samples prepared by two different ways are shown in Fig. 1. No XRD peaks corresponding to the cobalt and nickel compounds appeared, even in the case of 10 mol.% Co^{2+} and Ni^{2+} addition. The reason might be explained as these transition metal ions were doped into the TiO_2 lattice and/or the very fine dispersions of transition metal oxide were formed on the surface of the titania, resulting in X-ray amorphous behaviour [11-16, 20]. All samples showed very broad diffraction peaks, indicating small crystallite sizes. All of Co^{2+} added samples consisted of the mixed phases of anatase and rutile. In contrast, 10 mol.% Ni^{2+} added sample by the 1-step method mainly consisted of anatase phase, while the peak intensity of rutile increased with the decrease in Ni^{2+} content, indicating the depression of the formation of rutile by Ni^{2+}. However, 10 mol.% Ni^{2+} added sample by the 2-step method consisted of the mixed phases of anatase and ruitile. This indicates that the phase transformation of metastable anatase to a thermodynamically stable rutile proceeded, since the hydrothermal treatment was conducted twice in 2-step method. The samples with mixed crystalline phases were expected to show superior photocatalytic activities to single phase samples, since the effective charge separation by the heterogeneous electron transfer can be expected as P25 TiO_2 [21,22].

The morphology of the sample was observed by TEM, and the images are shown in Fig. 2. It is observed that the $TiO_{2-x}N_y$ consisted of mainly rod-like nanoparticles with the diameter below 10 nm and the length about 20 nm. The samples prepared by the 1-step method showed very similar morphology to that of $TiO_{2-x}N_y$ without doping modification, indicating that transition metal doping did not show any influence on the morphology. The samples prepared by the 2-step method showed larger rod-like particles, probably due to the Ostwald ripening under repeated hydrothermal treatment. Compared with the particle size of Fe^{3+} added $TiO_{2-x}N_y$ samples (range from 20 to 60 nm) prepared by the hydrothermal reaction using an electric oven in a previous work [12], Co^{2+} and Ni^{2+} added $TiO_{2-x}N_y$ prepared in the present study showed much smaller particle size. The smaller particle size might be attributed to the quick microwave-assisted hydrothermal reaction used in the present study.

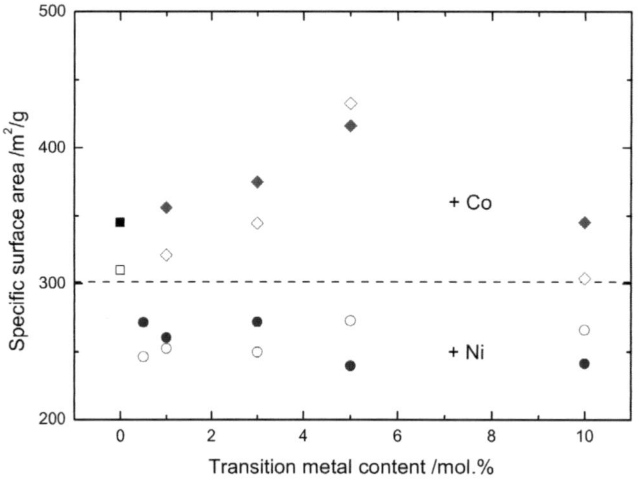

Figure 3. Specific surface areas of $TiO_{2-x}N_y$, and the samples prepared by 1-step or 2-step method with different Co/Ni content. (Solid dots: 1-step method; hollow dots: 2-step method)

The specific surface areas of the samples are summarized in Fig. 3. The specific surface areas of Fe^{3+} added samples fabricated using an electric oven were ca. 165 m^2/g [12], while those of Co^{2+} and Ni^{2+} added samples using a microwave oven were much higher as ca. 250-400 m^2/g. The values of 2-step samples were slightly smaller than those of 1-step samples, due to the crystal growth by twice hydrothermal reaction. It is also notable that the specific surface area did not change so much by adding Co^{2+}, but decreased by adding Ni^{2+}, indicating the promotion of the crystal growth in the presence of Ni^{2+}.

The diffuse reflectance spectra of Co^{2+} and Ni^{2+} added samples are shown in Fig. 4. By adding Co^{2+} and Ni^{2+} to $TiO_{2-x}N_y$, the visible-light absorption ability increased, especially above 500 nm of wavelength. For the samples prepared by the 1-step method, the enhancement in visible-light absorption by Co or Ni co-doping might be attributed to the formation of extra bands below the conduction band by mixing Ti 3d and Co or Ni 3d orbitals [11-16]. On the other hand, for the samples prepared by the 2-step method, the coupling of transition metal oxides with $TiO_{2-x}N_y$ effectively increased the visible light absorption in Fig. 5(C, D).

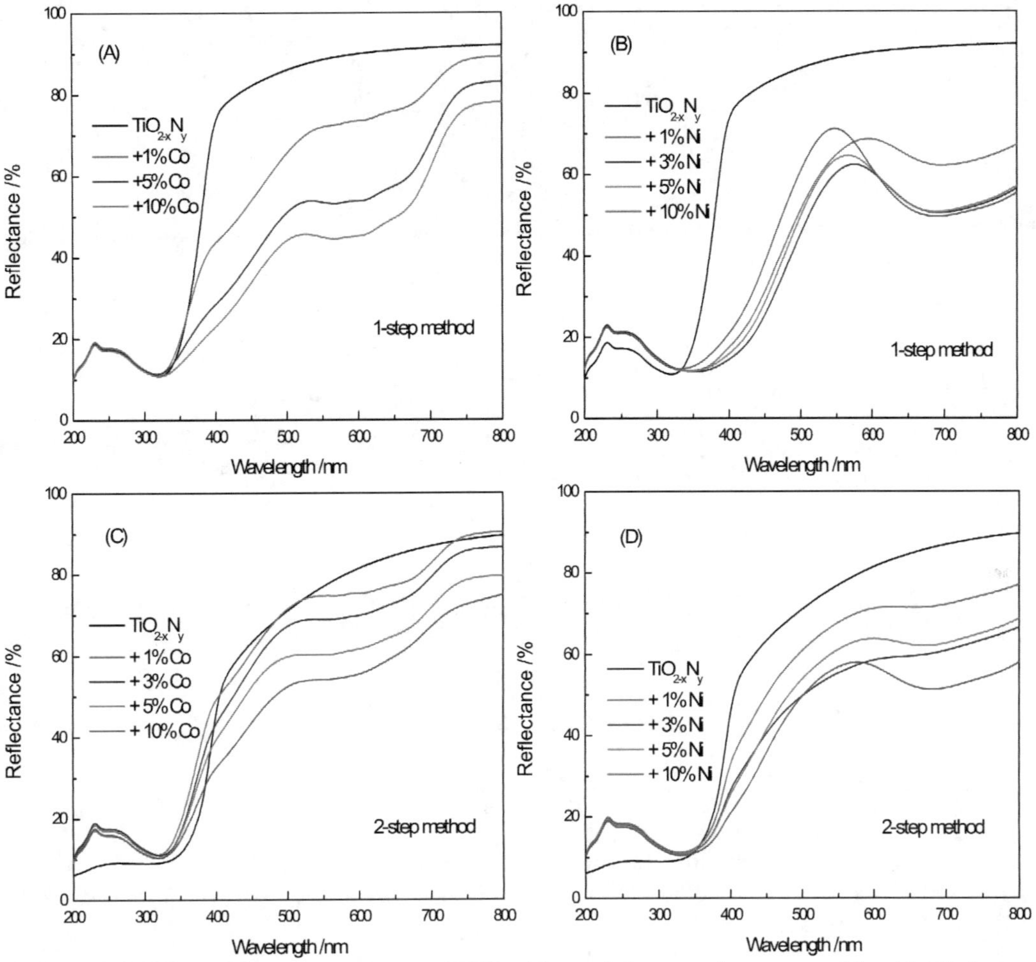

Figure 4. Diffuse reflectance spectra of $TiO_{2-x}N_y$, and the samples prepared by (A-B) 1-step method or (C-D) 2-step method with different (A,C) Co or (B,D) Ni contents.

The photocatalytic activities for the oxidative decomposition of NO gas are shown in Fig.5. The $TiO_{2-x}N_y$ without transition metal ion showed excellent activity even under visible-light irradiation due to the narrowing of band gap energy by N doping. In case of 1-step method, the addition of Co^{2+} resulted in the decrease in the photocatalytic activity, whereas the photocatalytic activity increased by the addition of a small amount of Ni^{2+} less than 5 mol.% and then decreased. The increase in the photocatalytic activity might be due to the increase in the visible-light absorption ability, whereas the decrease in the photocatalytic activity might be due to the promotion of the recombination of photoinduced electrons and holes by the increase in the oxygen vacancy for the charge compensation. Namely, the samples prepared by the 1-step method actually gave a co-doping of transition metal and N in TiO_2, in which transition metal ions substituted Ti^{4+} and N^{3-} substituted O^{2-} in the lattice structure. Anion vacancies would be formed in the lattice to maintain the electro-neutrality, which would act as the recombination centers for photo-induced electrons and holes to depress the activity. Bechstein et al. [23] also found the experimental evidence for enhanced surface oxygen vacancy formation upon Cr^{3+} doping of titania. Therefore, the more transition metal ions were co-doped into the products, the poorer the photocatalytic activity of the samples would be, due to the increasing amount of anion vacancy.

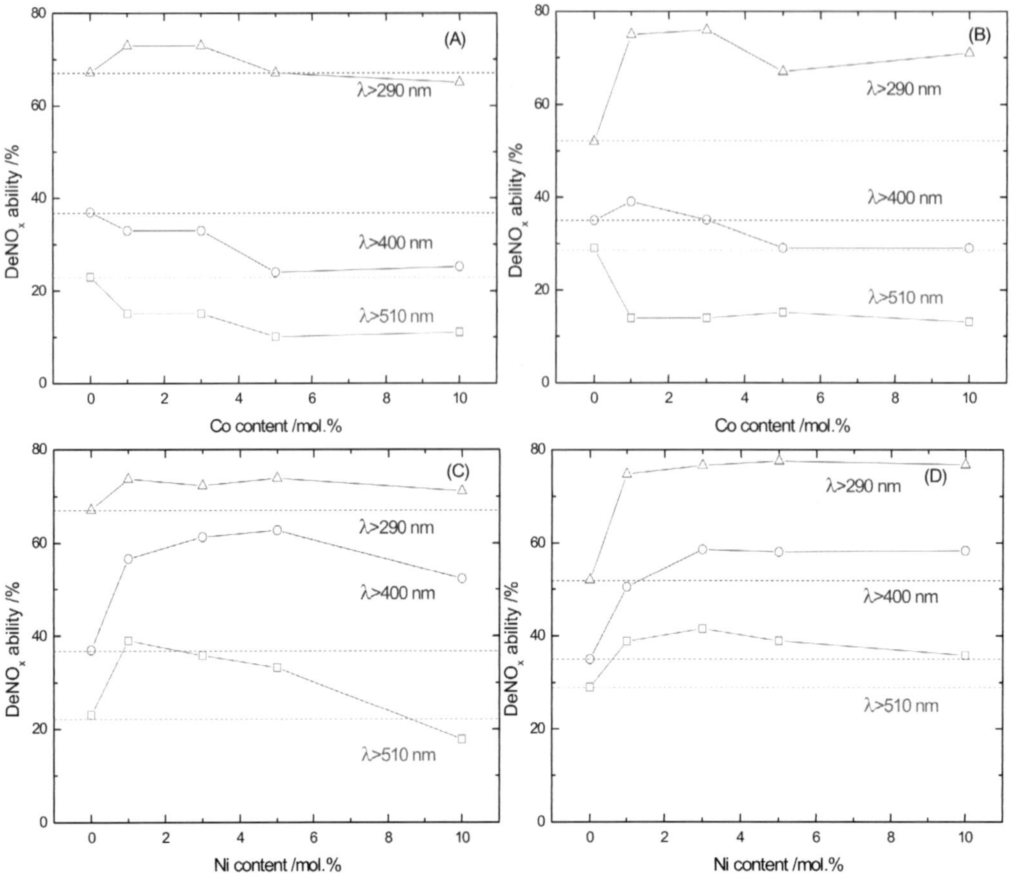

Figure 5. Photocatalytic activities for the oxidative decomposition of NO of $TiO_{2-x}N_y$, and the samples prepared by 1-step method (A, C), or by 2-step method (B, D), with different Co (A, B) or Ni (C, D) contents.

On the other hand, the photocatalytic activity of Ni^{2+} added sample by the 2-step method increased with an increase of Ni^{2+} content up to 5 mol.% and then was almost constant. In case of the 2-step method, $TiO_{2-x}N_y$ particles were first precipitated, followed by the precipitation of NiO particles on the surface of $TiO_{2-x}N_y$ particles to form $TiO_{2-x}N_y$/NiO composites. Therefore, the increase in the photocatalytic activity in the initial stage may be due to the effective separation of photo-induced electrons and holes by the heterogeneous electron transfer from $TiO_{2-x}N_y$ to NiO. Even though the samples prepared by the 2-step method showed lower specific surface areas, they still showed higher photocatalytic activities than those prepared by the 1-step method. Co^{2+} added sample by the 2-step method also showed higher photocatalytic activity than that by the 1-step method, but the enhancement by Co^{2+} addition was modest and the addition of an excess amount of Co^{2+} resulted in the decrease in the photocatalytic activity, although the reason has not been clarified yet.

The photocatalytic behaviours of Ni^{2+} added $TiO_{2-x}N_y$ were similar to that of Fe^{3+} added $TiO_{2-x}N_y$ reported in a previous paper, but the activity of Ni^{2+} added sample prepared in the present study was higher than Fe^{3+} added samples. The formation of mixed phases of anatase and rutile phase under this 2-step synthesis method could also favor high photocatalytic activity. Furthermore, Ni-modified $TiO_{2-x}N_y$ showed optimal photocatalytic deNO$_x$ abilities under both visible-light and UV light irradiation.

4. Conclusion

With the same starting materials, transition metal (Co or Ni) ion-modified $TiO_{2-x}N_y$ was synthesized by 1-step and 2-step hydrothermal method. Transition metal ion doping and transition metal oxide loading could be synthesized by these two methods. The photocatalytic activity in NO destruction was decreased by the co-doping of excess amount of transition metal ion, due to the increase in oxygen vacancy. In contrast, the combination of $TiO_{2-x}N_y$ with transition metal oxides by the 2-step hydrothermal method showed a positive effect to enhance the photocatalytic activity owing to the effective separation of photo-induced electrons and holes. The current results suggested the importance of element status on the photocatalytic activity, and also provided a clue to design highly active photocatalyst.

Acknowledgments

This research was supported in part by the Management Expenses Grants for National Universities Corporations from the Ministry of Education, Culture, Sports, Science for Technology of Japan (MEXT), and by the Grant-in-Aid for Science Research (No.20360293 & No. 22651022).

References

[1] Mills A and Hunte S L 1997 *J. Photochem. Photobiol. A* **108** 1.
[2] Hashimoto K, Irie H and Fujishima A 2005 *Jpn. J. Appl. Phys.* **44** 8269.
[3] Choi W, Termin A and Hoffmann M R 1994 *J. Phys. Chem.* **98** 13669.
[4] Anpo M and Takeuchi M 2003 *J. Catal.* **216** 505.
[5] Bellardita M, Addamo M, Di Paola A and Palmisano L 2007 *Chem. Phys.* **339** 94.
[6] Di Valentin C, Pacchioni G and Selloni A 2004 *Phys. Rev. B* **70** 085116.
[7] Reddy K M, Baruwati B, Jayalakshmi M, Rao M M and Manorama S V 2005 *J. Solid Satate Chem.* **178** 3352.
[8] Zhao W, Ma M, Chen C, Zhao J and Shuai Z 2004 *J. Am. Chem. Soc.* **126** 4782.
[9] Asahi R, Morikawa T, Ohwaki T, Aoki K and Taga Y 2001 *Science* **293** 269.
[10] Morikawa T, Asahi R, Ohwaki T, Aoki K and Taga Y 2001 *Jpn. J. Appl. Phys.* **40** 561.
[11] Yu J G, Xiang Q J and Zhou M H 2009 *Appl. Catal. B* **90** 595.
[12] Zhang P, Yin S and Sato T 2011 *Appl. Catal. B* **103** 462.
[13] Shen S, Chen J, Wang X, Zhao L and Guo L 2011 *J. Power Sources* **196** 10112.
[14] Ghasemi S, Rahimnejad S, Rahman Setayesh S, Rohani S and Gholami M R 2009 *J. Hazard. Mater.* **172** 1573.
[15] Hameed A, Gondal M A and Yamani Z M 2004 *Catal. Commun.* **5** 715.
[16] Zhou B, Zhao X, Liu H, Qu J and Huang C P 2011 *Sep. Purif. Technol.* **77** 275.
[17] Li H H, Yin S and Sato T 2011 *Appl. Catal. B* **106** 586.
[18] Yin S, Hasegawa H, Maeda D, Ishitsuka M and Sato T 2004 *J. Photochem. Photobiol. A* **163** 1.
[19] Anpo M 2002 Recent Development on Visible Light Response Type Photocatalyst *NTS, Tokyo* **ISBN 4-86043-009-03** p.9.
[20] Ambrus Z, Balazs N, Alapi T, Wittmann G, Sipos P, Dombi A and Mogyorosi K 2008 *Appl. Catal. B* **81** 27.
[21] Hurum D, Agrios A and Gray K 2003 *J. Phys. Chem. B* **107** 4545.
[22] Ohno T, Tokieda K, Higashida S and Matsumura M 2003 *Appl. Catal. A* **244** 383.
[23] Bechstein R, Kitta M, Schutte J, Kuhnle A and Onishi H 2009 *J. Phys. Chem. C* **113** 3277.
[24] Li H H, Yin S and Sato T 2011 *Nanoscale Res. Lett.* **6** 5.

AUTHOR INDEX

Chang, Y. K. 55
Dong, Q. 14, 60
Dong, Y. 9
Fan, L. 43
Fukumori, A. 18
Guo, C. 60
Guo, C. S. 14
Hsieh, J. H. 55
Huang, G. 9
Huang, Y. 9, 33
Ioku, K. 5
Kamitakahara, M. 5
Kinomura, N. 14
Kitazawa, K. 1
Kojima, Y. 1
Kubota, A. 18
Kumada, N. 14
Li, C. 55
Li, H. 14, 60
Li, Y. 47
Lin, J. 33
Liu, C. X. 37
Liu, G. 47
Luo, T. 47
Nishimiya, N. 1
Saito, T. 5
Sato, T. 14, 60
Sato, Y. 18, 28
Shiota, T. 24
Takei, T. 14
Tokoro, Y. 28
Uemura, K. 24
Wang, L. 9
Wang, Y. 60
Watanabe, T. 28
Wei, Y. 33, 43
Wu, J. 33, 43
Xiang, L. 37
Xie, G. 43
Xie, Y. 33
Xu, L. 47
Yao, X. 47
Yasuda, K. 24
Yin, S. 14, 60
Yonesaki, Y. 14
Yoshikado, S. 18, 28
Yu, R. B. 55
Zhang, F. 9
Zhang, L. Z. 37
Zhang, P. 60
Zhang, X. 9